WIND RESOURCE ASSESSMENT AND MICRO-SITING

WIND RESOURCE ASSESSMENT AND MICRO-SITING
SCIENCE AND ENGINEERING

Matthew Huaiquan Zhang

机械工业出版社
CHINA MACHINE PRESS

Contents

Preface

I joined Vestas China in early 2008 when China was undergoing a tremendous boom in the wind energy industry with the installed capacity of wind power more than doubling year on year. It was indeed an exciting time, but very soon I realised that the industry was literally running with blind eyes. Wind data were measured carelessly. Projects were executed with merely a few months' poor wind data and being built where power grids were already saturated with wind energy. The number of wind turbine manufacturers went from a few to close to a hundred in just a few years! Wind resource assessment and turbine siting was considered unimportant and had to very often give way to the top-to-bottom bureaucratic planning of a speedy development. I have therefore felt the urge to bring sense back to the industry ever since I started in the industry. At the time, there was no book in the Chinese market on this subject and the wind engineers were generally equipped with insufficient knowledge to deliver a sound wind resource assessment, owing at least partly to the low requirements from the industry. I then decided to write one myself. After a year of hard work, the first edition of the book was finally published by China Machine Press in June 2013. The market has responded very positively.

I have to admit that writing a technical book while keeping a full-time job is quite tough. It basically means no weekends, no holidays and working until midnight nearly every day for a whole year. It was all worth it in the end but I certainly wished that I would never put myself into the same kind of stress ever again. Yet, as soon as the Chinese edition was published, I pretty much had to immediately come to terms with writing it in English all over again. It all happened because my partner and I made the decision of relocating to the UK, but doing so meant leaving 10 years of life in Beijing behind. I therefore got in contact with John Wiley & Sons, Ltd in the autumn of 2013 in order to pursue the publication of the book in English, and managed to come to an agreement with them. As a second-language speaker who only learnt the language at school, writing it in English seemed overly ambitious. This time, though, I resigned from my full-time job and became an independent consultant so that I could gain a better control of my time as well as working on the process of my relocation to the United Kingdom. Instead of directly translating the content into English, I have taken the opportunity to refine the areas where I did not have time to do so in the Chinese edition and also catering to an international audience. Consequently, the book you are reading now is essentially the second edition with sizable improvements and reshuffling of the first one.

My educational background, an MSc in Material Science and second major in International Economy, has little to do with the domain of wind energy application. Also, before joining Vestas, I worked on website development. It may seem a bit sporadic but it also makes the book interesting because it is composed by a hands-on specialist who first approached the subject as a total stranger and gained a high level of proficiency. The structure and content of the book are carefully selected and designed on the basis of my own learning experience of the subject. The domain of wind resource assessment and micro-siting involves multiple disciplines ranging from statistics and flow modelling to atmospheric physics and meteorology. Each chapter in this book can be a much specialised area of study, easily filling its own books. This may be one of the reasons why this subject has rarely been organised into one structured and coherent text. Therefore, the content in this book does not intend to be exhaustive or overly precise. Many aspects of wind resource assessment are still developing and undergoing sometimes heated debates, such as numerical wind flow modelling and uncertainty analysis. I, myself, am still learning new things and gaining more experience every day. The aim of this book is therefore not to decisively settle every debate, but I do hope that it can provide the readers with sufficient information and insight to make sound decisions in each critical step of wind resource assessment and in delivering more wind power projects that perform effectively. For those who wish to further their knowledge in any specific topics or enter academics, this book can still serve as a gateway or a good foundation.

Matthew Huaiquan Zhang

Introduction

The successful development of wind energy projects depends on an accurate assessment of the wind conditions and siting of each wind turbine. Wind in nature is a very complex meteorological phenomenon with a dominant feature of constant and sometimes violent fluctuations. From measured wind data, which often only covers a time range of a couple of years, to the estimate of 20 years of average wind power production of the wind farm, it takes multiple complicated analyses with inherited risks. A mistake in this stage of evaluation can cause severe financial losses and missed opportunities for developers, lenders, and investors.

Wind Resource and Micro-siting, Science and Engineering constructively and coherently pulls together all the key theories of the domain, aiming to form a strong, systematic foundation of wind resource assessment and micro-siting for readers. It should allow readers to utilise the contents for research purposes as well as a go-to guide for useful information. The areas covered in this book include analytical and numerial wind flow modelling, wind statistics, wind measurement techniques and data analysis, MCP, uncertainty analysis, wind energy meteorology, offshore micro-siting and environmental impact assessment. In order to assist readers' learning, most topics start with easy-to-understand background details and ease into the practical demonstration of day-to-day engineering work. The author brings his own experience to bear on the teaching and applications of his knowledge acquired over the years, which makes it reader friendly. It aims to build a bridge between general professionals, through to more advanced and specialist researchers in each topic.

Acknowledgments

I would like to thank Vestas Wind System A/S, the organisation that brought me into the domain of wind resource assessment and provided me with extensive training from the very beginning. A few topics are inspired by and built upon what I learnt in the training programmes, for example wind data analysis and uncertainty analysis delivered by Kim E. Andersen and meso-scale meteorology by Line Gilstad. I am also grateful for the permissions to use and reproduce many of the figures and contents in the book by organisations including Risø National Laboratory, DTU, IEA Wind R&D and EMD International A/S, and the support from both China Machine Press and John Wiley & Sons, Ltd. Notwithstanding their diligent efforts and support, any errors and oversights remain the sole responsibility of the author.

Matthew Huaiquan Zhang

About the Author

Receiving his Masters degree in 2006 from Tsinghua University, China, the author entered the wind energy industry in 2008. He began his career as a wind and site engineer with Vestas, the world's leading wind turbine manufacturer, gaining WAsP certification in 2009 in a training and examination programme in Chicago, USA. Matthew Huaiquan Zhang has dedicated himself to gaining expert knowledge in all aspects of wind resource related technologies. Over the years, he has successfully assessed over 5 GW of wind park capacity and trained a team of competent engineers for the Chinese wind energy market.

List of Symbols

a	scale parameter in Weibull and generalised Pareto distribution
	fractional wind speed reduction along the rotor surface normal
	slope parameter of a straight line
	acceleration
b	scale parameter of a straight line
b_j	threshold at neuron j (ANNs)
d	displacement height
e	residual in regression models
f	frequency
	nomalised frequency in Equation (8.18)
	Coriolis parameter
$f(x)$	probability density function
	regression function
g	Earth's gravity
h	height
h_H	wind turbine hub height
k	shape parameter in Weibull, extreme value and Pareto distribution
	entrainment constant in Jensen wake model
	kurtosis
	factor in Equation (5.89)
l	height where the maximum speed-up effect is found
m	mass
n	number of records
	frequency in Equation (8.18)
p	number of fitted parameters in Equation (6.8)
q	specific humidity (mixing ratio) of the air mass
r	radius
s_c	mean crosswind relative spacings in a regular grid layout
s_d	mean downwind relative spacings in a regular grid layout
s_i	relative distance to the neighbouring turbine i
t	time (scale)
u	horizontal wind speed
	turbulence fluctuation component in horizontal direction
u'	turbulence term of horizontal component

u_*	friction velocity
u_0	horizontal wind speed at the top of the hill
v	wind speed
	perpendicular wind speed
	turbulence fluctuation component in perpendicular direction
v'	turbulence term of perpendicular
v_i	ith wind speed record in a wind data series
w	turbulence fluctuation component in vertical direction
	vertical wind speed
w'	turbulence term of vertical component
w_j	weight of jth data point
x	distance to the roughness change line
	horizontal axis of an orthogonal coordinate system
	a random variable
y	perpendicular axis of an orthogonal coordinate system
z	height level above surface
	vertical coordinates
z_r	reference height level
z_0	roughness length
z_{00}	wind-farm equivalent roughness length
AEP	annual energy production
A_W	section area of turbine wake
	location A in WAsP model
B	location B in WAsP model
Ce	electrical power coefficient
CF	capacity factor
C_p	power coefficient
C_T	thrust coefficient
D	rotor diameter
D_{eff}	effective rotor diameter
D_i	Cook's distance
D_W	diameter of wind turbine wake
E	operator of mean
E_{PF}	energy pattern factor in Equation (5.60)
Err	prediction error
F	force
F_c	centrifugal force
F_g	gravitational force
F_f	friction force
F_j	cumulative probability of ranked data point i (Gumbel method)
F_r	Coriolis force
	Froude number
F_p	pressure gradient force
F_T	cumulative probability of a T year period
$F(x)$	cumulative distribution function
G	gust factor
I	turbulence intensity

I_0	ambient turbulence intensity
I_{eff}	effective turbulence intensity
I_{ref}	average turbulence intensity at wind speed of 15 m/s
I_u	streamwise (horizontal) turbulence intensity
I_T	maximum turbulence intensity in the centre of turbine wake
I_W	wake added turbulence intensity
K	kinetic energy
K_m	turbulent transfer coefficient
L	half-width of the middle of the hill
	loss in annual energy production
	Monin–Obukhov length
L_p	sound pressure level (dB) at the receptor
L_w	sound power level from a noise source (dB)
LAE	least absolute error
MSE	mean square error
N	number of neighbouring wind turbines
	number of records
	static atmospheric stability parameter
N_M	number of full years of onsite wind measurement
N_P	financial horizon of a wind power project
N_{ref}	number of years of the reference dataset (MCP)
N_{target}	number of years of the concurrent dataset at target site (MCP)
ORO	effects of orography in WAsP model
OBS	effects of obstacle in WAsP model
P	air pressure
	wind power density
	probability
	sound power intensity
P_e	electrical power of a wind turbine generator
P_{free}	energy production at free-stream wind speeds
P_{park}	predicted energy production of a wind park (wake effect included)
P_r	air pressure at a reference level z_r
P_w	wake probability
P95	annual energy production with 95% probability of exceedance
P90	annual energy production with 90% probability of exceedance
P75	annual energy production with 75% probability of exceedance
P50	annual energy production with 50% probability of exceedance
R	specific gas constant for air
	distance from a sound source
R^2	coefficient of determination
	correlation coefficient
Re	Reynolds number
RIX	ruggedness index
RMSE	root mean square error
ROU	effects of roughness (WAsP model)
S	sensitivity (production uncertainty)
	upstream cross-section of roughness element A section area

	turbine swept area/rotor area
$S(n)$	spectral power density
SS_{err}	sum of squares of error/residual
SS_{tot}	total sum of squares
SSR	sum of squared errors
T	recurrence period
	temperature
T_H	highest monthly-average temperature
T_L	lowest monthly-average temperature
T_v	virtual temperature
Tr	air temperature at a reference level z_r
TKE	turbulence kinetic energy
U	mean wind vector component in horizontal direction
U_0	free stream horizontal mean wind speed
U_g	geographic wind vector
U_H	free stream hub height wind speed
U_H'	wake-affected hub height wind speed
Up	predicted wind speed
U_M	measured wind speed
V	mean wind vector in perpendicular direction Volumn
V_{50}	50-year extreme mean wind speed
V_{ave}	annual mean wind speed at hub height
V_{gust}	gust wind speed
$V50,_{gust}$	50-year extreme gust wind speed
V_i^M	median value of wind speed bin i
V_{ref}	reference wind speed in IEC standard
V_T	extreme wind speed of recurrence period T
W	mean wind vector in vertical direction
W_{reg}	common regional wind climate (WAsP model)
W_A	micro-scale wind climate of location A (WAsP model)
W_B	micro-scale wind climate of location B (WAsP model)
X	a variable
	distance
$\overline{X}$	mean term of any meteorological variable
X'	fluctuation term of any meteorological variable
$\widehat{Y}$	predicted value of Y (MCP)
α	wind shear component
	constant in Equation (4.40)
	scale parameter in generalised extreme value distribution
	Charnock parameter
	frequency-dependent sound absorption coefficient
β	initial wake expansion parameter
	location parameter in generalised extreme value distribution
γ	skewness
λ	crossing rate in Equation (5.88)
σ	standard deviation/uncertainty

σ_1	intrinsic uncertainty of the 1-year mean wind speed
σ_{ann}	standard deviation of the annual mean wind speeds
$\sigma_{conbined}$	combined uncertainty
σ_i	uncertainty component
σ_H	standard deviation of average high tempeture
σ_L	standard deviation of average low tempeture
σ_N	intrinsic uncertainty of the N-year mean wind speed
σ_P	uncertainty in long-term windiness
σ_u	standard deviation of streamwise (horizontal) velocity
σ_v	sample standard deviation of wind speed
$\sigma_{v,hh}$	uncertainty of wind speed at hub height due to vertical extrapolation
σ_V	standard deviation of mean wind speed
θ	potential temperature
θ_i	ith wind direction record in radius
θ_i^o	ith wind direction record in degree
θ_v	virtual potential temperature
τ_w	surface shear stress
ρ	air density
μ	mean of a random variable
μ_n	nth central moment
μ_v	sample mean wind speed
μ_x	easting mean of directional components
μ_y	northing mean of directional components
μ_θ	sample mean wind direction in degrees
ρ_i	ith air density record in a wind data series
κ	Kármán constant
	turbulent kinetic energy (TKE)
ε	rate of dissipation of turbulence kinetic energy
	spread of wind direction data points
η_e	electrical efficiency of a wind turbine generator
η_m	mechanical efficiency of a wind turbine generator
η_{Park}	wind park efficiency
ξ	location parameter in generalised Pareto distribution
ζ	stabliliy parameter
ω	angular velocity of the earth's rotation
ω_{ij}	weight on the input from neuron i to j (ANNs)
φ	latitude
Ψm	stability function
Γ	gamma function
$\Delta\alpha$	uncertainty in wind shear
Δu	relative horizontal fractional speed-up (flow over hills)
ΔRIX	RIX difference between the predicted and reference sites
∇p	differential pressure gradient

1

Introduction

Energy supply is undoubtedly one of the most challenging issues facing human beings in the 21st century. Limited traditional fossil fuels are being used up gradually, let alone air pollutants and global warming caused by the combustion of those dirty fuels. Renewable energy has therefore attracted increasingly more attention in recent years. Wind energy, being one of the most commercially viable forms of renewable energy at the moment, has already played an important role in quenching our society's energy thirst. Yet there is still a long way to go before we can fully exploit the wind potential of our planet.

Wind is the 'fuel' for wind power generation. Its characteristics, that is wind conditions, are therefore of the upmost importance when it comes to determine the economics of a wind farm project. Wind conditions are set by nature, but how well we can understand or estimate them is another question. Wind resource assessment in essence is the estimation of wind conditions based on wind data available and topographical (roughness, obstacles and terrain) and meteorological (e.g. atmospheric stability, boundary layer structure, weather system) features of a given site.

Being invisible already makes it hard for us to picture the wind, and to make matters worse it varies constantly and dramatically in time and space, influenced by a great number of factors, some of which we may not even know of. However, on the other hand, building wind farms is very capital intensive and those wind farms have to generate profit for their owners. Profitability has to be predicted before wind farms are built with a reasonable risk premium. Such stringent requirements from the industry have raised sometimes almost impossible challenges for wind resource assessment professionals. After all, the results of wind resource assessment and micro-siting will determine the success of the investment of a wind power project.

This book endeavours to bring together pieces of core knowledge used in wind resource assessment and to put them into a logical order and to explain them, adding in the author's own experience obtained in day-to-day work scenarios. This kind of effort has rarely been made before, at least to the author's knowledge, even though a few publications covering a few sections of the domain can be found in the market.

Wind Resource Assessment and Micro-siting, Science and Engineering, First Edition. Matthew Huaiquan Zhang.
© 2015 John Wiley & Sons, Ltd. Published 2015 by John Wiley & Sons, Ltd.

1.1   Wind Resource Assessment as a Discipline

From a meteorological point of view, the study of a wind resource for the purpose of energy production can be described as wind energy meteorology, which has developed into an independent division of meteorology. In fact, a monograph named *Wind Energy Meteorology* by Emeis [1] has recently been published in early 2013, a milestone of the discipline. Petersen *et al.* [2] describe wind energy meteorology as applied geophysical and fluid dynamics, a combination of meteorology and applied climatology.

Despite its importance, wind energy meteorology has not been a major area of expertise required by the industry to produce satisfactory wind resource analysis results until the last decade or so. In the last decade especially, wind turbines have substantially grown in size and height, which means that they are exposed to much more complicated atmospheric boundary layer structures. Simplified engineering models, which worked well before, have to be re-examined based on the study of wind energy meteorology. The fact that wind turbines are usually erected in more complex terrain conditions, and even offshore nowadays, has also promoted the development of the discipline. Therefore a significant portion of time will be spent on this subject in order to form a physical profile of wind resource analysis for readers.

Wind resource assessment takes us one step closer to the wind energy industry, setting off from the ivory tower of physics. The domain of wind resource assessment should at least consist of wind data analysis, site analysis, wind turbine selection, wind turbine siting (micro-siting), wind flow modelling, power production estimates, wind park optimization and uncertainly analysis. Statistical tools are predominantly used in the process owing to the stochastic nature of the wind. Therefore, statistics becomes another pillar of wind resource analysis, the first one being the physical models explained by wind energy meteorology, such as the boundary layer profile and atmospheric stability. In order to ensure quality calculations, we need to understand how the wind should be measured as well as interaction mechanisms between wind and wind turbines and amongst turbines (wake effects).

The development of wind resource assessment has been accompanied and motivated by the commercial evolution of wind turbines and the construction of large-scale wind power projects. It will continue to do so in the foreseeable future. As a matter of fact, the expertise in wind resource assessment has become a core competence for many organizations in the industry and therefore well sought after.

1.2   Micro-siting Briefing

Micro-siting is really a meteorological definition, because in the eyes of a meteorologist, a few hundred metres is really on a micro scale. Micro-siting can be defined as the process of strategically positioning wind turbines within a given project area, in order to maximise power production with minimised turbine loads, that is optimising the wind park. Petersen *et al.* give an alternative definition of micro-siting, that is an estimation of the mean power produced by a specific wind turbine at one or more specific locations [2].

A full siting procedure includes considerations such as the availability and capacity of the power grid, the present and future land use, and so on, but these aspects are not considered in this book. However, one important issue concerning the siting of wind turbines is their environmental and health impact, such as noise and flickering, which can turn into a dominant factor in some cases and is explicated in Chapter 11.

1.3 Cascade of Wind Regime

The wind in nature almost never travels along a straight line; rather its track resembles circles. Those 'circles' are of all sizes, driven or dominated by different forces and induced by various mechanisms. Bigger 'circles' break into smaller ones and then even smaller ones until dissipated into heat, that is vibration of air molecules.

The wind we feel is a superposition of all the 'circles' of air movement at one spot. The scale of wind regime (or the size of the 'circles') can be described by two dimensions: temporal and spatial. The temporal scale and spatial scale of a wind regime are closely related. We can imagine that the bigger it is in space, the longer it takes to finish a circle. This cascade of wind regime should be the first physical model of the wind one should formulate before getting into the world of wind resource assessment. Wind regime is also referred to as wind climate or wind system. Chapter 9 will present wind systems of various scales in detail.

1.3.1 Global Scale Wind Regime

The atmosphere is a very complex heat engine whose energy is supplied by the heating of the earth's surface by the sun. Because the earth is tilted and also because of its uneven surface, different parts of the earth receive substantially different amounts of energy from the sun, which in turn induces air circulations with a spatial scale of the entire globe and a temporal scale of one or many years. This partly explains why wind resources are distributed so unevenly around the globe, as shown in Figure 1.1 [2].

Long-term wind data measured around the globe are required to analyse wind climate on this scale, but such efforts are commonly hindered by poor data quality (usually measured at

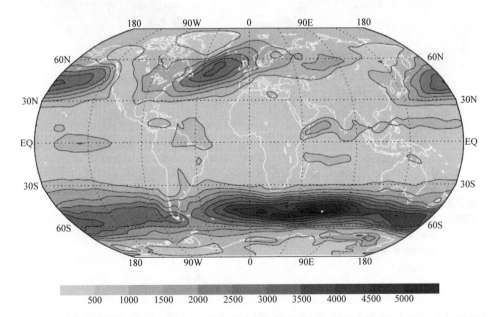

Figure 1.1 Energy flux of the wind at 850 hPa (about 1500 m.a.s.l.) in W/m^2 from 8 years of the NCEP/NCAR reanalysis [2] *Source*: Risø National Laboratory

10 m height and contaminated by local features and inconsistent through time) and insufficient measurement points.

In recent years, however, the advances in computational power, the availability of nontraditional meteorological datasets with global coverage (such as satellite data), in addition to the traditional ones used in the global meteorological network (e.g. the Global Observing System [3]), and the advances in weather prediction models have together made it possible to reconstruct the global scale weather situation at every instant over recent decades. Global meteorological models are able to provide dynamic, consistent wind data and statistics, while avoiding some of the setbacks associated with the direct use of wind data (in fact most reanalyses do not consider low-level wind data in the analysis because of their 'contamination' with local influences) [4]. Figure 1.1 [2] is a good example of such applications and indicates the global wind resource variation, though the figure is rather dated. Figure 1.2 demonstrates the distribution of wind power density in China.

Global meteorological models are usually made with spatial resolutions too coarse to be used in project-based wind resource analysis. For a higher level of spatial resolution, which includes smaller-scale phenomena of significant influence on the wind resource, it is now

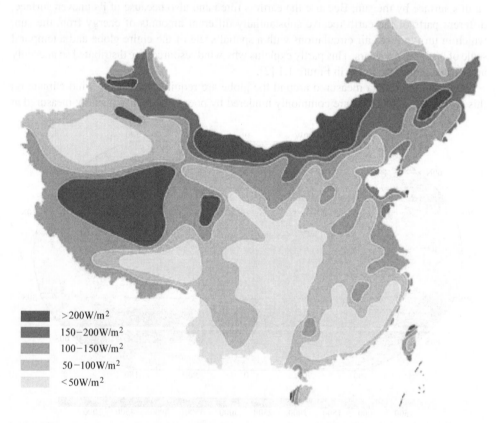

>200W/m²
150–200W/m²
100–150W/m²
50–100W/m²
<50W/m²

Figure 1.2   Wind power density distribution of China *Source*: China Meteorological Administration (CMA)

a common practice to use mesoscale NWP models of much higher resolutions than global reanalysis models. These models cover a smaller domain using global data as boundary conditions. From the aspect of wind resource assessment and micro-siting for wind power projects, this global scale wind climate is a rather remote application, and therefore is generally excluded in this book. However, it does indicate the importance of long-term correction of in-situ wind measurements, which typically cover only a period of one or two years.

1.3.2   Synoptic Scale Wind Regime

We are probably more familiar with the synoptic scale wind regime because of the application of weather forecasts, which frequently mention weather systems, such as fronts, pressure systems, cyclones and anticyclones. Synoptic scale wind regimes usually have a spatial scale of more than 200 km (up to thousands of kilometres) and temporal scales of over a few days (up to a few months). Large synoptic scale wind systems (e.g. westerly winds in mid-latitude, trade winds and monsoons) are in most cases the main drivers of a local wind resource, although often severely altered by smaller scale phenomena.

Synoptic scale systems are obviously also too large for the purpose of analysing wind resource conditions within a wind farm and wind turbine siting in spite of its significant impact on local winds. The analysis of synoptic scale systems for wind turbine micro-siting only becomes relatively more important in some offshore cases where the overall atmospheric stability (one of the determining factors for wind turbine spacing and layout design) is closely related to wind direction and the passage of weather systems. Even then, it is usually not studied in detail. This is not to undermine the importance of understanding synoptic scale wind systems for wind resource assessment and micro-siting; rather it is to find the right focus for professionals working in this area as well as the scope of this book.

1.3.3   Meso-scale Wind Regime

Large scale models are insufficient to capture all weather phenomena related to the local wind resource. Meso-scale models zoom in from such large scale models and look at the missing scales between large scales and local or micro-scales. It covers special wind phenomena such as land and sea breezes, tropical hurricanes and convective storms. In the meantime, the meso-scale is also commonly used to explicit large scale flows modified by terrain features such as hills, mountains and surface features, which are not resolved in the larger scale models. As large scale and global models become more and more sophisticated and of increasing resolution, the division between large scale and meso-scale becomes blurred.

Meso-scale NWP models are commonly used nowadays in wind resource assessments in terms of a site hunt, that is searching for possible sites for wind power projects. In contrast with micro-siting, which deals with the wind resource within a given project area, a site hunt using meso-scale models can be referred to as macro-siting. Meso-scale NWP models are also able to generate a long-term wind data series, which is invaluable for long-term correlation of on-site wind measurements. As a result, a wind resource analyst should know by heart the correct usage of the results from meso-scale models and their pros and cons (see Chapter 3).

1.3.4   Local Scale Wind Regime

The local wind regime reflects local wind effects down to the smallest scales superimposing on larger scale features and meteorological systems. From a meteorological point of view, it generally represents micro-scale meteorological features and wind effects, giving the origin of the term 'micro-siting'.

Local winds are driven by local geographical features. Near the surface, medium to small hills usually have similar levels of impact on wind speed with vegetation and obstacles [5]. Figure 1.3 illustrates the slightly exaggerated variation of mean wind speed at 10 metres above ground level on a typical Danish site, indicating clearly the effects of topographical features on local winds. During micro-siting, various micro-scale topographical features have to be carefully taken into consideration.

In general, in situ wind data measured by local meteorological masts (often referred to as a met mast in short) should be sufficient for the estimation of a wind resource at the measured point. However, it is more likely that we do not have measurement data at the specific point(s) of interest (at the hub height of the intended wind turbine at their precise locations) or not over a sufficiently long time (years) to be able to directly estimate the wind resource. Therefore, it is essential to have an understanding of the most important mechanisms that influence the wind locally. Micro-scale differs from meso-scale and larger scales in that their influence in general can be assessed with respect to mean winds and wind resources by corrections based on empirical relationships and simple physical models [4].

Local influences that must be considered include sheltering by obstacle(s) (more important at relatively low heights), speed-up effects of orography (e.g. hills, valleys), roughness and surface thermal conditions (or atmospheric stability, which is especially important for assessing the wind resource at higher heights above the surface). Chapter 2 will elaborate on these effects in terms of analytical models and engineering applications, while Chapter 3 is dedicated

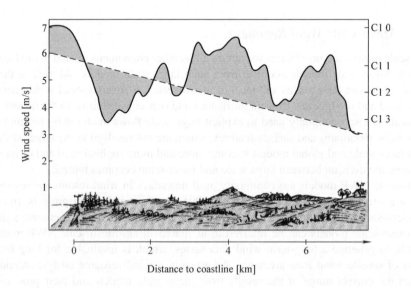

Distance to coastline [km]

Figure 1.3   The variation (slightly exaggerated) of mean wind speed 10 m.a.g.l. due to topographical effects (full line) for typical conditions of Denmark [5] *Source*: Risø National Laboratory

to numerical wind flow modelling. The more theoretical roots of those effects will be presented later in Chapter 9 and Chapter 10.

In most areas of the world, the local wind resource is essentially determined by large synoptic scale wind systems (they determine the overall long-term trend of a wind resource) with smaller scale features superimposed on them. The study of large scale wind systems requires long-term wind measurement, whereas local in situ wind measurements can be as short as merely one year. Therefore, long-term correlation of wind data is to some extent the study of the correlation between local effects and large scale systems. Chapter 6 will elaborate on this topic in detail.

Wind resource is a statistic quantity. Thus, the assessment of wind resource starts with the statistics of local wind speed and direction. In an overwhelming majority of cases, an on-site met mast(s) must be installed in order to reduce uncertainties of the wind resource assessment and turbine load calculations. The number of necessary masts depends on the complexity of the site. The measurement height should be as close to the expected wind turbine hub height as possible. Measuring the wind at multiple heights is necessary in order to evaluate local influences on the wind better. Due to the difficulty of constructing met masts in many cases (e.g. offshore), ground-based remote sensing (lidar and sodar) capable of measuring wind up to hundreds of metres high have been deployed more widely in recent years. Wind measuring techniques will be introduced in Chapter 8.

1.4   Uncertainty of Wind Resource

The complexity of wind resource assessment to a great extent is due to the uncertainty of the wind in nature and the uncertainty of the tools we use to predict it. It is thus something we should always keep in mind. The characteristics of statistics itself also imply that the wind, as a statistical quantity of stochastic variables, must bring in various intrinsic uncertainties as well.

The random nature of wind resources makes wind energy commercially unique to traditional fossil fuels. The price of fossil fuels fluctuates substantially with the market, which is risky for the business, whereas wind is free and never changes in price. Therefore the revenue generated by a wind power plant is not affected much by the unpredictable energy market; instead it tends to grow in the long run as electricity price generally follows an upward trend over years. On the other hand, the risk of investing in wind energy lies on how well we can estimate the wind resource and calculate the power production before erecting wind turbines and investing real money. This explains the significant role wind resource analysis and micro-siting plays for delivering a commercially successful wind power project.

To help us understand the uncertainty of wind resource better, let us take a look at a well-known example of wind speed variations at a point in Ireland, as shown in Figure 1.4. The line of monthly mean wind speed in the upper frame in Figure 1.4 indicates violent fluctuations. Yearly mean wind speeds calm down significantly, but still vary dramatically from year to year, shown more clearly by the slash blocks in the lower frame in Figure 1.4. Even averaged over 10 years, the means still do not seem to be stabilised. Noting the fact that the energy content of wind is proportional to the third power of the mean wind speed, one can imagine the magnitude of the variation in available wind energy due to fluctuating wind conditions. For example, a 1% decrease in the mean wind speed can be expected to yield about 2% less energy.

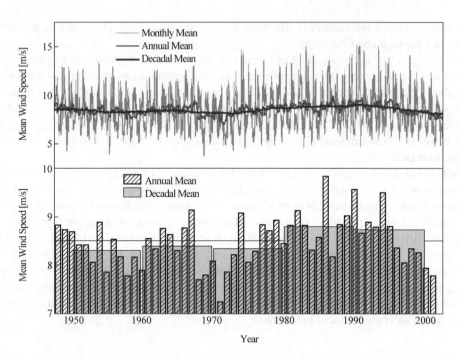

Figure 1.4 Reanalysis data for a point in Ireland showing the variability of the mean wind speed *Source*: Risø National Laboratory, created by Gregor Giebel

Unfortunately, we typically have only one or two years of in situ wind measurements, representing only wind conditions in the past. Then how can we accurately assess the mean wind speed over the next 20 years (the commonly expected life time of wind turbines) and predict power production and the profit of the wind power project for its entire life cycle? As we can see from Figure 1.4, the annual mean wind speed in 1971 was only about 7.2 m/s whereas it hiked to approximately 9.5 m/s in 1986. The peaks and troughs of annual wind climates not only make the profitability assessment of wind power investment a risky business but they may also cause a wrong selection of wind turbine types. According to the mean wind speed in 1971, this should be an IEC III site, but the data in 1986 points to IEC I, although the wind in the long term should actually be IEC II [6]. If IEC I wind turbines were chosen, the efficiency of the wind farm would be dampened. If IEC III turbines were chosen, the wind loads could be unbearable for the turbines, which would increase wear and tear and maintenance costs and even shorten the machines' lifetime.

Consequently, long-term correlation and correction of in situ wind data based on long-term reference data become extremely important. A method often used to reduce the uncertainty resulting from measurements over a short period is the so-called Measure–Correlate–Predict (MCP) method, which uses the statistical relation established between the wind data in a long term 'reference' dataset and a shorter time overlapping dataset measured at a 'target' site to estimate the target site wind statistics over the long reference period [3,7]. This methodology of MCP will be presented in Chapter 6, while the uncertainty study of wind resource assessment is placed later in Chapter 7.

1.5   Scope of the Book

As put by EWEA in the report *Wind Energy – The Facts* in 2009 [7], 'The wind is the fuel for the wind power station. Small changes in wind speed produce greater changes in the commercial value of a wind farm . . . . Commercial evaluation of a wind farm is required, and robust estimates must be provided to support investment and financing decisions . . . . Once the wind speed on the site has been estimated, it is then vital to make an accurate and reliable estimate of the resulting energy production from a wind farm that might be built there.' This generally summarizes the standpoint of this book.

Wind resource assessment ranges from synoptic scale to micro-scale. It is inclined more towards scientific meteorology for larger scale wind resource assessment. This book, on the other hand, focuses on micro-scale analysis and its engineering applications, providing practical and theoretical guidance for wind resource assessment and micro-siting. The theories presented in the book are also designed for a better conduct of day-to-day engineering work. Each chapter of the book can easily fill its own book, but the scope of this book is to sufficiently elaborate each topic for engineers and establish a platform for those who wish to dig deep in research. A professional working in this field does not have to be a specialist in meteorology or a statistician, but the knowledge put together in this book is critical and should be acquired.

References

[1] Emeis, S. (2013) *Wind Energy Meteorology*, Springer Verlag, Berlin and Heidelberg.
[2] Petersen, E.L, Mortensen, N.G, Landberg, L., Højstrup, J., Frank, H.P. (1997) *Wind Power Meteorology*, Risø National Laboratory, Denmark.
[3] Global Observing System. World Meteorological Organization, http://www.wmo.int (last accessed on 1 October 2013).
[4] Petersen, E.L., Troen, I. (2012) Wind Conditions and Resource Assessment. *WIREs Energy Environment*, **1**, 206–217.
[5] Troen, I., Petersen, E.L. (1989) *European Wind Atlas*, Risø National Laboratory, Denmark.
[6] IEC 61400-1 (2005) Ed. 3 – Wind Turbine – Part 1: Design Requirements.
[7] European Wind Energy Association (2009) Wind Energy – The Facts: Chapter 1, Technology. http://www.wind-energy-the-facts.org (last accessed on 1 October 2013).

2

Concepts and Analytical Tools

The scope of wind resource assessment ranges from an overall wind resource analysis of a large region, that is regional assessment, to a power production estimate for a single given wind turbine, which is achieved through micro-siting. Though micro-siting requires substantially more detailed information, they both utilise basic analytical concepts of topography and local wind regime. Topography is usually referred to as the whole variation in surface properties and elevation; orography, on the other hand, is used especially to address height elevation.

To address the critical influence of topography on wind, we will need to systematically describe the features of topography that are influential to wind. Three components of those features are orography or terrain, roughness and obstacles. Obstacles as a type of topographical feature are somewhere between orography and roughness and the lines are not that clear-cut. The interaction between topography and wind is extremely complex and heavily influenced by many other nontopographical factors, such as air density and thermal stratification of the atmosphere. Ideally the three components should be considered as a whole, but in reality the model will become too complex and fragile to solve. Therefore, as a trade-off, they are in most cases considered separately and in parallel.

Not only are they solved more easily by computers but simple analytical models can also help us to better understand the wind and make sound technical decisions during micro-siting. As simplifications of real case scenarios, these engineering applications should be able to equip new engineers to carry out many types of works after a short period of training.

This chapter will focus on addressing basic conceptual and analytical wind flow models. The classic content in this chapter has been covered by many publications, most famously the *European Wind Atlas* [1] published in 1981, which the author has no regret in referencing in large portions.

2.1 Surface Roughness and Wind Profile

2.1.1 Roughness Length

The surface of the earth (land and water) imposes friction on wind blowing on top of it. Roughness length is a measurement of the magnitude of surface roughness. It is generally determined

Wind Resource Assessment and Micro-siting, Science and Engineering, First Edition. Matthew Huaiquan Zhang.
© 2015 John Wiley & Sons, Ltd. Published 2015 by John Wiley & Sons, Ltd.

by the size and distribution of roughness elements the surface contains. For land surfaces, these elements are typically vegetation, built-up areas and the soil surface [1].

Physically, z_0 is independent of wind speed, atmospheric stability and stress, z_0 changes only if the roughness element of the surface changes, such as height and coverage of vegetation, urban development and deforestation, and z_0 increases but with a magnitude always smaller than the height of the roughness length.

Lettau (1969) [2] has presented a simple empirical relation between the roughness element and roughness length:

$$z_0 = 0.5\frac{hS}{A} \tag{2.1}$$

where h and S are the height and upstream cross-section of the roughness element respectively and A is the average horizontal area.

Equation (2.1) gives reasonable estimates of z_0 when roughness elements are evenly and spaciously distributed (A is much larger than S). If roughness elements are close together, the flow will be 'lifted' over them, creating a displacement height. Porosity of the roughness elements, such as trees and bushes, must also be considered when calculating the cross-section S.

Example 2.1

Small bushes are evenly distributed across a land area of $5000\,\text{m}^2$, the average height of each bush is approximately $50\,\text{cm}$ and its width is roughly $1\,\text{m}$. There are about 500 such bushes in the entire area. Due to summer, the bushes are well foliaged and therefore have a porosity of 20%. What is the estimated roughness length of the land area?

Solution

For each bush, as a roughness element, its cross-section S facing the wind is

$$S = 1 \times 0.5 \times (1 - 20\%) = 0.4\,\text{m}^2$$

The available horizontal area A for each roughness element is

$$A = 5000/500 = 10\,\text{m}^2$$

Then the estimated roughness length is

$$z_0 = 0.5\frac{hS}{A} = 0.5 \times \frac{0.5 \times 0.4}{10} = 0.01\,\text{m}$$

In reality, the land surface is often covered by roughness elements of various sizes and shapes. Kondo and Yamazawa (1986) [3] therefore proposed a similar empirical relation in order to calculate the roughness length in such circumstances:

$$z_0 = \frac{0.25}{S_T}\sum_{i=1}^{n} h_i S_i \tag{2.2}$$

where S_i is the cross-section of the ith roughness element facing the wind, h_i its height and S_T the total area of all roughness lengths.

The water surface is more complicated due to the dynamics of waves, which depends on the wind speed itself. Though the roughness length of the water surface is not constant, it is always very small. For this reason, the roughness length of the water surface is usually assumed constant for simplicity in wind resource assessment; for example 0.0002 m is assumed in WAsP. Such simplification is usually unacceptable when it comes to simulating turbulence on open water surfaces.

Roughness lengths are usually not calculated directly from the above-mentioned equations during real day-to-day work of wind resource assessment and micro-siting. Those equations are elaborated to give readers a better understanding of the influencing factors and physical characteristics of roughness length. Rather, roughness lengths are in most cases empirically estimated according to the features of roughness elements of the surface.

According to observations over a long period of time from various locations, the typical roughness lengths z_0 of different surface types are summarised in Table 2.1 [1].

As we may have noticed, the roughness length may vary significantly with seasons. For example, a farmland in the north (the northern hemisphere) may be covered by ice and snow in winter, giving a very small roughness length of 0.001 m, whereas in summer when crops and vegetation flourish and the roughness length becomes 0.05–0.10 m, there is an over 50 times increase compared to winter. Consequently, the only realistic option is to use the annual average value. It is obviously difficult to accurately estimate the annual mean value, even though theoretically this can be done by a weighted average of each season.

The simplest and necessary procedure is by cross-extrapolating/interpolating wind speeds measured at different heights from the same met mast based on the initially estimated roughness lengths. If the extrapolated or interpolated value does not match the measured one, then adjustments should be made accordingly until the results are satisfactory.

The concept of mean values is integrated into the entire process of wind resource assessment, such as mean air density, mean atmospheric stability, mean wind speed, etc. Due to the complexity of meteorological issues, where almost all of the parameters change constantly with substantial randomness, such engineering simplifications become very necessary.

Table 2.1 Roughness length and class for typical surface characteristics [1]

Terrain surface characteristics	Roughness length z_0 (m)	Roughness class
Forest and urban areas	0.7 ~ 1.0	
Suburbs and sheltering belts	0.3 ~ 0.5	3
Farmland with closed appearance, many trees, bushes	0.1	2
Farmland with open appearance, very few buildings, trees, etc.	0.02 ~ 0.05	1
Mown grass and airport runway areas	0.01	
Smooth snow surface	0.001	
Smooth sand surface	0.0003	
Water area	0.0002	0

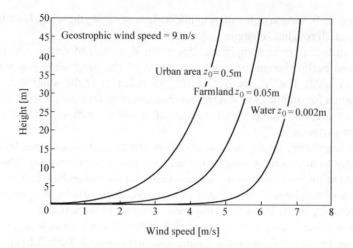

Figure 2.1   Idealised wind profiles for three different surface types

2.1.2   Vertical Wind Profile

At the geometric length where the lower boundary of the atmosphere occurs, the roughness length is the height where the mean wind speed in the logarithmic profile disappears. For a thermally neutral atmosphere and flat terrain, the vertical wind profile in the surface layer of the atmosphere is

$$\frac{u(z)}{u_*} = \frac{1}{\kappa} \ln\left(\frac{z}{z_0}\right) \tag{2.3}$$

where z_0 is the roughness length in metres, non-negative constant u_* is the friction velocity and κ is the Kármán constant (≈ 0.4); u_* square has the property of turbulent shearing stress. According to Equation (2.3), u_* and z_0 can be easily deduced if provided with mean wind speeds of two heights. The height where the mean wind speed equals zero can also be estimated by interpolating the logarithmic equation.

Surface roughness fundamentally determines the wind profile above; other factors such as terrain and atmospheric stability can be taken as adjustments and superposition over the effect of surface roughness for the sake of engineering simplification. The vertical wind profiles of three distinguishing surface roughness conditions in homogeneous topography and identical atmospheric stability are demonstrated in Figure 2.1.

Geostrophic winds can be considered as the wind atop where the surface frictions vanish. In Figure 2.1, the geostrophic wind is always 9 m/s, meaning that the three wind profiles would cross at the same point if extended upwards. Different roughness length result in substantially different wind speeds near the ground. An urban area with the roughest surface ($z_0 = 0.5$ m) slows the wind the most dramatically and presents the lowest wind speed near the ground. A water surface, on the contrary, causes the slightest slowdown of the wind speed due to its small roughness length (0.0002 m), indicated by the steepest wind profile. The three wind profiles all follow a logarithmic law nevertheless.

It seems obvious why wind speed offshore is usually higher than onshore. Possible changes of surface roughness during the lifetime of a wind farm, such as trees growing or urban development, must be noted and taken into consideration during micro-siting in order to better estimate the energy production of the wind farm.

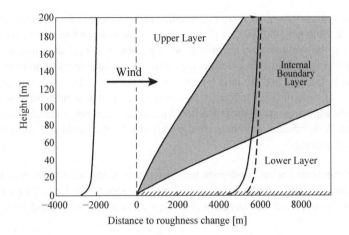

Figure 2.2   Sketch of a wind profile change after passing through a coastline that separates two distinctive roughness surfaces, sea on the left side of the dashed line and land on the right. Arrows represent the wind direction

2.1.3   Internal Boundary Layer

What happens to a wind profile if the roughness length changes suddenly, for example when wind passes a coastline that separates the land and the sea?

In Figure 2.2, the sea surface with a roughness length of 0.0002 m is on the left-hand side of the dashed line and land surface (roughness length 0.2 m) on the right, indicating a sudden roughness change. Before the wind reaches the coastline (sea breeze), the wind profile is very steep due to the low roughness length of the sea surface. The wind senses the change in roughness from the bottom as soon as it crosses the coastline, whereas the wind above remains unchanged and still is the sea breeze with a steep wind profile as before. As the wind pushes further inland, the impact imposed by the land surface grows in height, formulating a new profile following the new roughness, that is the wind turns into a land breeze. While the very top is still a sea breeze, there must be a transition layer between the upper layer and lower layer, which is the so-called internal boundary layer (IBL). The wind within the IBL is a mixture of sea breeze and land breeze and is the result of the interaction between two distinctive roughness lengths.

It is noteworthy that similar IBLs can also occur when wind passes through two underlying surfaces with different heat flux features (atmospheric stability), which is called the thermal internal boundary layer (TIBL). Coastlines again can be a very good example of this case.

The IBL is an important concept for wind resource assessment and micro-siting with many implications:

- *Roughness in distance is more influential.* Roughness at the foot of a wind turbine may be insignificant, because its influence is not yet felt by the wind at the hub height or even the bottom of the rotor. For instance, vegetation on mountain tops where wind turbines are usually sited is often low (low roughness) due to the high wind speed and lack of water, whereas surrounded by a vast forest with much higher roughness. What the wind turbines situated on such mountain ridges feel is actually the roughness of the forest rather than the grass at their feet. It must be taken into consideration when describing wind farm roughness.

- *The distance to roughness change is critical.* Taking Figure 2.2 as an example, if a wind turbine with a hub height of 80 m is placed onshore 200 m away from the coastline, the wind conditions at the wind turbine are still generally a sea breeze; if the same wind turbine is placed 8000 m away from the coastline, then traces of sea breeze will tend to disappear from the wind at hub height and the land breeze starts to dominate. It is therefore confirmed that the distance of wind turbines to roughness change is critical as wind conditions may change significantly, especially when the prevailing wind direction is perpendicular to the roughness change line. A significant roughness change may also happen on land, for example between forest and grassland.

The above-mentioned wind conditions have effects on many aspects such as wind profile, atmospheric stability and turbulence intensity. This issue should be carefully investigated when it comes to designing wind measurement campaigns, that is siting for a met mast for a certain wind farm.

- *Minimum Size of Roughness Map.* Since roughness features in the distance have critical impacts, then how big is big enough for a roughness map to be able to cover all the influential features suitable for wind resource assessment? Engineers like the idea of rule of thumb because they are practical and easy to use, but they are not necessarily robust standards per se. In Figure 2.2, the lower layer is approximately in the shape of a right triangle of which the ratio of the two right-angled sides is roughly 1:100. In other words, should the radius of the roughness map be larger than 100 times the wind turbine hub height, we can then say that generally all effective roughness features are covered. For a wind farm, the direct distance between the boundaries of the roughness map and the outermost wind turbines should be at least 100 times the hub height. Common wind turbines in the market nowadays usually have a hub height of 60–80 m so given the size of a wind farm area the radius of a roughness map should be at least 10 km.

2.1.4 Roughness Change Model

Roughness change usually comes along with the IBL phenomenon. The dramatic and clear-cut roughness change mentioned in the above section allows a clearer demonstration on this issue. We still use Figure 2.2 to showcase the roughness change scenario: wind travelling from roughness z_{01} to z_{02}. Starting from the roughness change line in the downwind direction, the wind speed above the IBL depends on z_{01} whereas the underneath IBL wind speed is adjusted by a parameter governed by the height of the lower layer h, the values of z_{01} and z_{02} and the distance to the roughness change line x (see Figure 2.3).

Then h can be expressed as [4, 5]:

$$\frac{h}{z_0'} \ln \left(\frac{h}{z_0'} - 1 \right) = 0.9 \frac{x}{z_0'}$$

$$z_0' = \max(z_{01}, z_{02}) \tag{2.4}$$

If hub height of a wind turbine is higher than h and the surface can be treated as homogeneous without considering roughness change. Otherwise roughness change has to be taken into account.

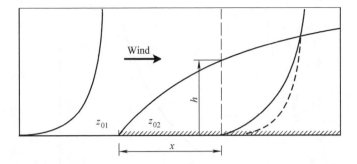

Figure 2.3   Idealised scenario of the wind profile before and after a roughness change

2.1.5   Displacement Height

For closely packed roughness elements, the surface level is essentially raised or 'displaced'. Dense forest canopies, for example, seem to have merged into one entity. A similar effect happens for heavily populated urban areas where buildings are tall and compact. As shown in Figure 2.4, when wind blows across a dense forest, the wind profile above the canopy still follows the same logarithmic law as Equation (2.3), equivalent to the ground being levelled up for a certain distance, referred to as the displacement height, i.e. d in Figure 2.4.

As the result of displacement height, only the top part of the canopy is reflected as the roughness element, rather than all the trees. The wind profile with the displacement height can easily be solved by adding a parameter into Equation (2.3), under the same neutral atmospheric stability:

$$\frac{u(z)}{u_*} = \frac{1}{k} \ln\left(\frac{z-d}{z_0}\right) \tag{2.5}$$

where d is the displacement height. It is not difficult to solve Equation (2.5) if the mean wind speeds of three different heights are known.

For wind turbines, the existence of displacement height is equivalent to a shortened hub height. If a wind farm is surrounded by a vast forest such as Yichun City in northeast China,

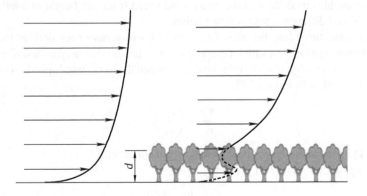

Figure 2.4   Illustration of wind profile blowing across a dense forest

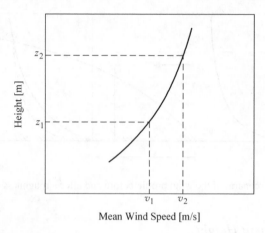

Figure 2.5   Indication of the mean wind speeds at two heights

the displacement height has to be taken into account because otherwise the wind speed at hub height and energy production of the wind farm would be severely overestimated.

The displacement height of a forest falls into a half to a third of the average height of the trees depending on the density and foliation of the trees. Cross-interpolations between wind speeds of different heights from the same mast should be conducted in order to verify the setups of displacement height and roughness length.

Trees are a very complicated issue for wind resource assessment and micro-siting, because they vary in size, shape, distribution and orientation, etc., and all shed direct impacts on the wind flows, making it almost impossible to come up with a simple model to give satisfactory simulations. Displacement height in some cases can be regarded as an alteration of the terrain or as obstacles.

2.1.6   Wind Shear

Wind shear is an alternative way of expressing wind profile, an engineering application really. Wind profile enables us to deduce the mean wind speed from one height to another and tells us the wind speed difference between two heights.

Instead of logarithmic law, Equation (2.3) or (2.5), which have been derived from physical and dimensional arguments, a simpler empirical power law is often used to describe the vertical wind profile, that is to describe the relationship between the mean wind speeds at two different heights (see v_1 and v_2 in Figure 2.5):

$$\frac{v_2}{v_1} = \left(\frac{z_2}{z_1}\right)^{\alpha}$$ (2.6)

where α is the wind shear component or wind shear for short.

By rearranging Equation (2.6), we get

$$\alpha = \frac{\ln(v_2/v_1)}{\ln(z_2/z_1)}$$ (2.7)

In the case of displacement height, z_2 and z_1 must be substituted by $z_2 - d$ and $z_1 - d$.

Wind shear α can be calculated as long as the mean wind speeds at two heights, v_1 and v_2 are known, which in turns enables the extrapolation of mean wind speeds at other interested heights, such as the hub height of a wind turbine.

Example 2.2

The annual mean wind speeds measured at 50 and 70 m.a.g.l. (metre above ground level) are 6.8 m/s and 7.3 m/s respectively. What is the mean wind speed at 60 m hub height?

Solution

Following Equation (2.7), the wind shear component can be calculated:

$$\alpha = \frac{\ln(7.3/6.8)}{\ln(70/50)} = 0.21$$

Let $z_1 = 70$ m and $z_2 = 60$ m; then $v_1 = 7.3$ m/s and Equation (2.6) will lead to the mean wind speed at the 60 m hub height:

$$v_2 = v_1 \left(\frac{z_2}{z_1}\right)^{\alpha} = 7.3 \times \left(\frac{60}{70}\right)^{0.21} = 7.07 \text{ m/s}$$

As an approximation, the power law wind profile is not without limits. Analysis [6, 7] has shown that only for certain conditions in stably stratified boundary-layer flow is it possible to find a power law profile that fits the logarithmic profile almost perfectly over a wide height range, although the fit becomes the better the smoother the surface is, that is the worst fit happens for unstable air conditions and rough terrain.

Another constraint of power law is that the power law Equation (2.6) may only be a good approximation to the real wind profile in the surface layer of the atmosphere. In fact, extrapolation of the wind profile above the height of the surface layer (80–100 m) by either logarithmic law (2.3) or (2.6) should be made very carefully because both laws are only valid for the surface layer only [8].

Due to the fact that the atmosphere is usually stably stratified in the mean [7] and that the extrapolation of mean wind speeds in a wind energy application is usually over a rather small height range (10–20 m), it is understandable why the power law has been successful in many cases. This fact is reiterated by the implication of mean throughout the process of wind resource assessment. Nevertheless, the limits of the power law profile and even the logarithmic law profile must not be overlooked.

Wind resource assessment software (e.g. WAsP and WindPRO) usually calculate wind shear from the wind profile derived from a roughness map. The resulting wind shear has to be verified by real wind data measured by a mast. If discrepancies occur, adjustment to the roughness map must be made until the simulated wind shear matches the measured one within a satisfactory margin. It also explains to some extent why it is important to measure wind speeds at a minimum of two different heights, because doing so is an indispensable way to minimise uncertainties in terms of vertical extrapolation of the wind speed.

Similar to wind profile, wind shear in real case scenarios is not static at all; rather it varies with mean wind speed, wind direction and atmospheric stability, etc. Most modern wind turbines have a cut-in wind speed of 3–4 m/s. Consequently, the wind speed data lower than the cut-in speed should be omitted from wind shear calculations.

From the curvature nature of the power law, we can conclude that the wind shear value also depends on height and the distance between the two heights and thus the greater the height difference the larger wind shear value. For a wind turbine, wind shear usually refers to the wind shear between the top and the bottom of the rotor plane.

2.2   Speed-up Effect of Terrain

Terrain effects are of paramount importance to the shaping of the wind flow near the ground and therefore is one of the key aspects of wind resource engineering.

The influences of terrain on the wind flows tend to decrease with height above ground level until they vanish entirely at the height where the wind blows horizontally without changing in speed. Therefore the wind flow crossing over a hill has to squeeze through a narrower pass (see Figure 2.6). Wind speed has to increase in order for the same amount of air mass to get through (mass-conservative), that is the terrain has a speed-up effect for the wind. Figure 2.6 depicts the speed-up effect by the density of the streamlines; the denser the streamlines the higher the wind speed, assuming attached flow near the surface, which is valid for gentle slops and simpler terrain.

Because of the speed-up effects, wind turbines are most likely sited on the summit in mountainous terrain. The speed-up effect is also subject to atmospheric stability, which will be elaborated in Chapter 10.

2.2.1   Horizontal Speed-up Profile

The effect of height variations in the terrain on the wind profile can be best demonstrated by the results from the international field experiments on the Wind Energy Conversion System at Askervein hill (116 m high) on the west coast of the Island of South Uist in the Outer Hebrides of Scotland (57°11′N, 7°22′W; UK Ordnance Survey grid reference: NF 754237). Askervein is

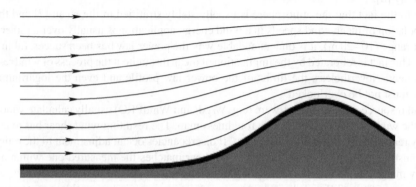

Figure 2.6   Streamlines of wind flow over an idealized hill

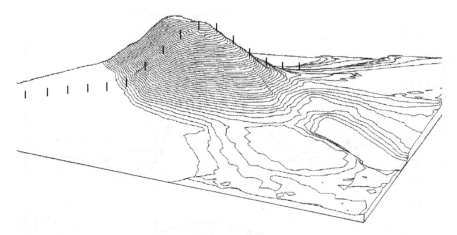

Figure 2.7 Perspective plot of Askervein hill and the layout of the masts [1] *Source*: Risø National Laboratory

a smooth hill with a quite simple geometrical shape resembling an ellipsoid (see Figure 2.7). Field experiments at Askervein were conducted in 1982 and 1983 with over 50 met masts deployed (10 m high, distributed in a straight line along the wind direction but perpendicular to the orientation of the ridge). The well-documented results have been used to verify and test many different wind flow models. For details of the project, readers can refer to an overview of the project presented by Taylor and Teunissen (1987) [9].

The relative horizontal fractional speed-up, Δu, can be defined as

$$\Delta u = \frac{u_0 - u}{u} \tag{2.8}$$

where u_0 is the horizontal wind speed at the top of the hill and u is the horizontal wind speed at the same height above ground level over the terrain upstream of the hill.

Many works (e.g. References [1,10] and [11]) have presented results comparing different wind flow models based on the Askervein hill project. Hyun *et al.* (2000) [11] have made comparisons of the horizontal speed-up profiles at a height of 10 m above ground level among the wind-tunnel experiments, full-scale measurements, two-dimensional computations, three-dimensional computations and theoretical predictions, as shown in Figure 2.8. Significant agreement among all results is found in Figure 2.8, particularly on the windward slope of the hill. Differences on the leeside can be considerably large and only the three-dimensional model catches this behaviour, although the agreement worsens at the peak. It is proposed that this phenomenon arises as the result of the three-dimensional flow separation on the lee slope caused by the influence of neighbouring hills [12]. If the neighbouring hills were excluded, flow separation on the leeside of the Askervein hill would not be predicted and would give similar results to those predicted by two-dimensional computation [11]. The fact that wind turbines are usually sited on top of mountain ridges means that the much more complicated and expensive three-dimensional wind flow models may not necessarily act better when it comes to predicting the wind speeds.

Some other noteworthy features from Figure 2.8 are that the horizontal speed-up at the top of the hill is about 80% as compared with undisturbed upstream mean wind speed, whereas a 20–40% negative speed-up effect occurs in the front and the lee of the hill.

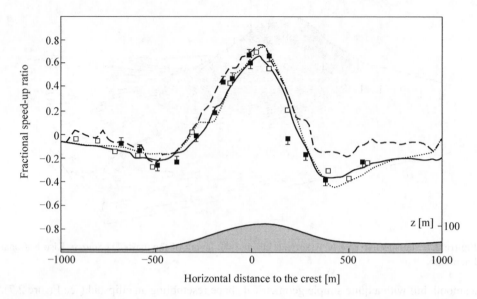

Figure 2.8 Horizontal fractional speed-up ratios for wind flow over Askervein hill at 10 m above ground level. Hollow rectangles: wind tunnel data; solid rectangles: field measurement; solid lines: three-dimensional numerical model; dotted lines: two-dimensional numerical model; dashed lines: theoretical model [11]

2.2.2   Vertical Speed-up on Hill Top

Once again, we use the Askervein hill project as the example. Figure 2.9 displays wind profiles recorded simultaneously upstream (undisturbed by the hill) and on the top of the hill.

From Figure 2.9, it is clear that the vertical wind speed-up reaches a maximum at the height of l, below and above which the wind profile follows a separate linear trend. Above l, however, the wind speed seems to remain constant with height. Extrapolating the upstream and hill-top wind profiles results in an intersection at the height of $2L$. The two characteristic length scales are defined in Figure 2.10, that is L is the half-width of the middle of the hill and l is the height where the maximum speed-up effect is found.

Jensen *et al.* (1984) [13] proposed an approximate expression for fractional speed-up Δu and l:

$$\Delta u \simeq 2\frac{h}{L} \tag{2.9}$$

$$l \simeq 0.3\, z_0 \left(\frac{L}{z_0}\right)^{0.67} \tag{2.10}$$

The formulas above usually work well when the dimension of the hill perpendicular to the wind direction is significantly larger than L, that is a long and gentle ridge, and the flow can be considered as two-dimensional. Figure 2.9 also validates linear wind flow models under similar terrain conditions. However, for a more complex terrain, it becomes difficult and sometimes impossible to accurately calculate the wind speed at a given site using Equations (2.9) and (2.10), and more sophisticated numerical models have to be employed.

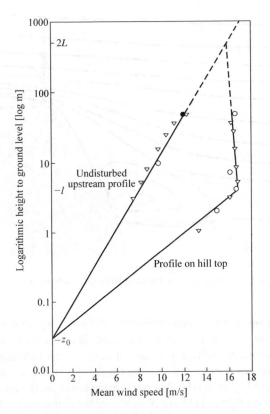

Figure 2.9 Simultaneously recorded undisturbed upstream wind profile and the profile on top of the Askervein hill (Jensen *et al.*, 1984 [13]). The symbols represent wind speed measurements. Both profiles correspond with those in Figure 2.10 *Source*: Risø National Laboratory

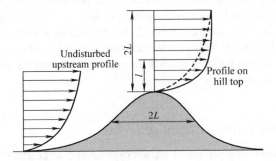

Figure 2.10 Flow over an idealized hill with undisturbed upstream and hill-top wind profiles. Two characteristic length scales are also indicated: L, the half-width of the middle of the hill, and l, the height of maximum relative speed-up [1] *Source*: Risø National Laboratory

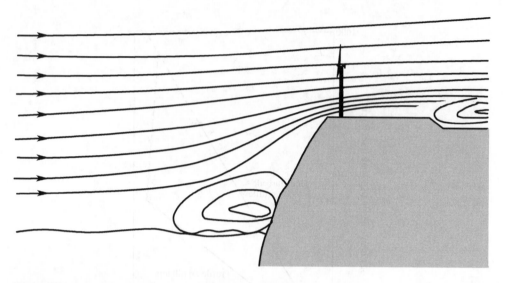

Figure 2.11   Wind flow over an escarpment blowing from the sea to land

The existence of the maximum speed-up height l in Figure 2.9 points out that the higher hub height (meaning higher investment) may not inevitably lead to a higher energy production on top of a hill as we would normally expect on flat terrain if the hub height is already higher than l. In fact, wind speeds may start decreasing (negative wind shear) above l for some more complex terrains where higher hub heights could mean lower energy production.

As Troen and Petersen (1989) [1] have pointed out, the definition of 'height above ground level' may not be a robust one in some cases. Imagine a wind turbine on a low hill: if the height of the hill is very small compared to the hub height and the sides slope steeply, the hill can be considered as a foundation for the turbine, which adds to the hub height, and the heights therefore should be counted from the hill base. However, if the dimensions of the hill are increased, the situation changes and the relevant height becomes the height above the hill top.

Troen and Petersen (1989) [1] also give another example: a wind turbine on top of an escarpment, commonly seen on a coast (see Figure 2.11). It would be wrong to use the height above the sea level as the hub height when the wind blows from the sea because the flow has been influenced by the cliff long before it reaches the land and therefore the relevant height should be the height above ground level.

2.2.3   Orographic Categorisation of Terrain

The terrain is the underlying surface of the atmosphere and its height variations exert a profound influence on wind, as elaborated in the above section. One of the main purposes of categorising terrain in terms of orography is to help us determine the suitability and uncertainty of different wind flow models. Figure 2.6 demonstrates attached (to the ground) flow or laminar flow, but for steeper slopes the flow tends to become detached and flow separation occurs (see Figure 2.12). Vortexes are generated in the separate flow layer, forming a virtual surface for the flow above, which remains by and large laminar. The slop of the virtual surface is gentler than

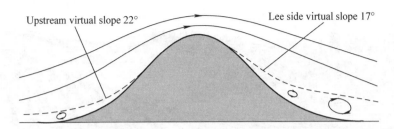

Figure 2.12 Virtual surface created by flow separation indicated by vortexes. Arrows represent wind direction; dashed line the virtual surface

the real slope of the hill. Separation that no longer follows linear laws profoundly complicates the issue of wind flow modelling and wind resource assessment [14, 15]. The likelihood of flow separation happening therefore becomes the ground for categorising terrain, from simple, that is no flow separation, to extremely complex, that is flow separation occurs.

It is generally accepted that flow separation occurs when the upstream slope is greater than 22° or the lee slope greater than 17° under neutrally stratified atmospheric conditions [14], such as the virtual slopes shown in Figure 2.12. Because the wind in nature may come from any direction, 17° is usually taken as the rule-of-thumb threshold between simple and complex terrain.

Sketches of the characteristics of different terrain types can be found in Troen and Petersen (1989) [1]; see Figures 2.13 to 2.16 for a practical qualitative classification.

Cat. I (Figure 2.13) is flat plains where winds near the surface are only modified by roughness changes and sheltering obstacles.

Cat II (Figure 2.14) is simple hilly landscape with gentle slopes (<17°). Typical horizontal dimensions of the hills are within a few kilometres. Winds near the surface are not only modified by roughness and obstacles but also accelerated by the hills. Attached flows can still be assumed and therefore linear wind flow models are generally sufficient for satisfactory simulation results.

Troen and Petersen (1989) [1] put broad and persistent sloping regions into a different category even though the slopes are within 17°, because, in such regions, meso-scale flow processes such as channelling, deflection and leeside descent may overrule local mechanisms and invalidate micro-scale flow models. This will be elaborated in Chapter 10.

Figure 2.13   Cat. I: Simple flat [1] *Source*: Risø National Laboratory

Figure 2.14   Cat. II: Simple hilly [1] *Source*: Risø National Laboratory

Figure 2.15   Cat. III: Complex hilly [1] *Source*: Risø National Laboratory

Cat. III (Figure 2.15) is complex hilly terrain. Once again, the typical horizontal dimensions of the hills are less than a few kilometres. This is where flow separation can no longer be assumed vacant and flow simulation becomes more complex with uncertainties too great to ignore. Linearised flow models tend to overestimate the orographic speed-up effect in these cases and three-dimensional flow models are usually required to simulate the winds. Even though linear models such as WAsP are incapable of accurately calculating winds in such complex terrain, some argue that corrections can be made in order to achieve more acceptable results (see Section 2.2.4) [16]. Miller and Davenport (1998) [17] point out that the speed-ups exerted by complex terrain are reduced in comparison with those found on isolated hills, as the effect of having a number of hills in series dulls the terrain effect, making the hills act almost like a very rough surface.

Cat. IV (Figure 2.16) is extremely complex mountainous terrain involving high mountains and deep valleys, which are outside the boundaries of most wind flow models, if not all. Thermally induced valley winds may dominate the wind climate [1]. The pattern of the atmospheric boundary layer may be altered as well. On the other hand, whether or not wind

Figure 2.16 Cat. IV: Extremely complex mountains [1] *Source*: Risø National Laboratory

power projects should be attempted in those terrain conditions deserves careful scrutiny in the first place.

One of the compounding issues of complex terrain is the turbulence generated by flow separations. Turbulence is notoriously harmful for wind turbines as it exponentially increases turbine loads and may induce severe wear and tear and increased breakdowns, a critical issue concerning the development of wind power projects in mountainous areas.

Figures 2.10 and 2.12 show idealised smooth hills, but the surface of real hills is rather rough. The flows in Figures 2.10 and 2.12 are stable, but real winds are turbulent. In reality, all descriptions of turbulent flow over natural land should ideally contemplate two very difficult modelling issues: surface roughness and flow separation. These two features usually happen simultaneously; therefore their effects should be considered together rather than separately. However, due to the excessive level of complexity in terms of describing both subjects, workers are usually forced to approach them individually [15]. Only in recent years have researchers been able to propose unified models; for example Loureiro *et al.* (2008) [8] proposed a new treatment for the lower boundary condition that, in principle, can be used in the regions of attached and separated flows over rough surfaces. However, those combined models are in general still not of common practice to date.

2.2.4 Ruggedness Index

The Ruggedness Index (RIX) can be considered as an effort to quantify the level of orographic complexity of the terrain and a coarse measure of the extent of flow separation [16]. Generally speaking, the higher the RIX value, the more complex the terrain and the more likely and strongly it is for flow separation to exuberate. RIX has been used extensively in wind resource assessment and siting studies in complex terrain if linearised flow models such as WAsP are applied. RIX of a given site is defined as the fractional extent of the surrounding terrain that is steeper than a certain critical slope [16, 19].

Niels *et al.* (2006) [16] have presented the calculation of RIX for a given site. In a polar coordinate system, each radius R originating from one site is divided into line segments defined

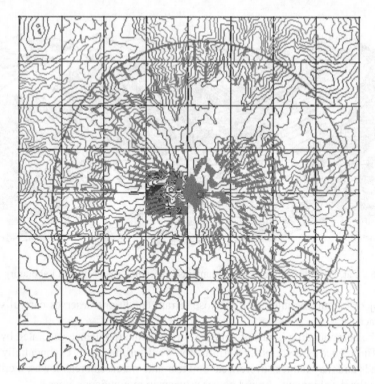

Figure 2.17   RIX calculated by WAsP Map Editor. Line segments with slopes greater than a certain threshold are indicated by the thick radial lines [16] *Source*: Risø National Laboratory

by the crossing of the radius with the height contour lines. The RIX value of the radius in question is the sum of the line segments representing slopes greater than a critical slope θ_c divided by the entire radius R. The overall RIX value for the site is then simply the mean of the radius-wise RIX values (see Figure 2.17).

The three determining parameters of RIX are radius R, the critical slope θ_c and the number of radii N, which are usually taken as 3.5 km, 0.3 (17°) (marking the onset of flow separation [14]) and 72 respectively (as shown in Figure 2.17), in compliance with the characteristics of the WAsP flow model [16, 19].

Attached flows are assumed when RIX ≈ 0, that is all slopes of the terrain are less than 0.3. If RIX > 0, it is indicative that somewhere in the surrounding area the slopes are greater than 0.3, which implicates flow separation.

The RIX value is regarded as an indicator of the adaptability of WAsP models in complex terrain, which will be elaborated further in the next chapter.

2.3   Shelter Effect of Obstacles

Obstacles to the wind such as buildings, trees and fences, etc., can reduce wind speed significantly and at the same time increase turbulence intensity in their vicinity. Both effects are not welcome to wind turbines.

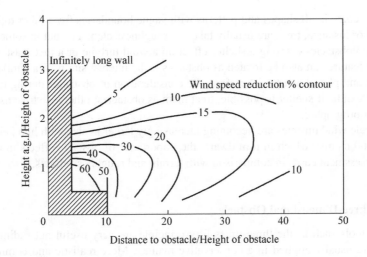

Figure 2.18   Percentage reduction of wind speed due to shelter by a two-dimensional obstacle [1]
Source: Risø National Laboratory

2.3.1   Reduced Wind Speed

The influence of an obstacle to a neighbouring site depends on: the geometrical shape, relative positing and orientation of the obstacle, the porosity of the obstacle and the height of the site.

2.3.1.1   Two-Dimensional Obstacle

Figure 2.18 shows shelter or shadow from an infinitely long two-dimensional obstacle of zero porosity, indicated in wind speed reduction as a percentage. As a rule of thumb, the shelter effect from such an obstacle disappears when the distance from the obstacle is larger than 50 times the height of the obstacle and/or the height of the site is taller than 3 times the height of the obstacle. This is the so-called three-times-fifty principle. It is a very conservative principle as obstacles are usually limited in length and may not be solid, thus allowing the air to breathe through.

Example 2.3

A met mast situated 60 m away from a 4 m high house measures the wind speeds at heights of 10 m, 30 m, 50 m and 70 m respectively. Which measurement heights are affected by shelter from the house?

Solution

According to the three-times-fifty principle the conservative shelter zone of the house is 12 m high and 200 m in distance. Therefore it is possible for the 10 m anemometer to be influenced by the shelter from the house whereas the 30 m, 50 m and 70 m anemometers are excused.

Obstacles come in all shapes and patterns with vague boundaries from other topographical elements. For instance, trees are usually taken as roughness elements, but in some cases they can turn into obstacles exerting a shelter effect on a wind turbine or a met mast. Hills as an orographic feature can also be treated as obstacles if the interest site is of a similar height to the hill top and placed nearby. Therefore, the consideration of obstacles during micro-siting should be flexible: if shelter is possible, treat them as obstacles; otherwise treat as roughness elements or orographic.

Utility scale wind turbines are becoming increasingly larger in size with hub heights easily exceeding 60 metres, which in turn dwarfs the importance of obstacle effects in many wind resource assessment cases in comparison with terrain and roughness effects.

2.3.1.2 Three-Dimensional Obstacle

Analytical tools such as the three-times-fifty principle are very useful but rudimentary and therefore are usually applied in a conservative manner. More realistic and commonly seen obstacles are three-dimensional, with comparable dimensions in length, width and height, such as a building or a tree.

The main differences in terms of wind speed distributions downstream of a three-dimensional obstacle in comparison with a two-dimensional one are: wind speed is accelerated on top of the obstacle, creating a speed-up zone and wind wraps in behind the obstacle from the sides supplementing the speed loss, meaning a milder speed reduction. The speed-up effect induced by an obstacle is similar to that of a hill and should be expected, but the difference is that the dimensions of an obstacle are significantly smaller than hills, thus giving a much smaller speed-up zone, too small to actually be utilized by utility scale wind turbines (this can be a positive aspect for urban small wind turbines on rooftops), let alone the dangerous turbulence (see the next section) that comes with it. From the wind speed reduction prospect, the influenced zones are more likely to be within 2 times the height of the obstacle in height and 10 to 20 times (depending on its girth and porosity) in distance downstream, which could be called a two-times-twenty principle (see Figure 2.19).

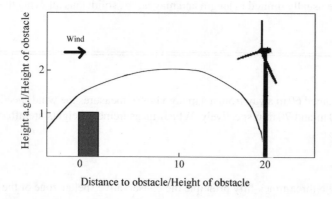

Figure 2.19   Sketch of the shelter zone by a three-dimensional obstacle

2.3.2   Increased Turbulence Intensity

Not only do obstacles reduce wind speed in their vicinity downstream but turbulence intensity is also significantly enhanced by shear forces exerted on the flow by the obstacle. Wind flow distortions around an obstacle are shown Figure 2.20, suggesting that the turbulent zone may extend to approximately 3 times the height of the obstacle [20] and is more pronounced behind the obstacle than in front of it [20].

Therefore, obstacles are generally the elements to avoid during wind turbine siting, especially when the obstacle is upstream of the wind turbine in the prevailing wind direction.

For a complex terrain, a small peak adjacent to a wind turbine may also be treated as an obstacle by using the two-times-twenty principle or the three-times-fifty principle should more conservative measures be necessary. It is also indicative from this perspective that complex mountainous areas eligible for wind park construction should at least have subdued and levelled ridges without pronounced variation in heights; otherwise the prominent peaks may generate unbearable turbulence intensity for wind turbines behind. If an obstacle-like peak is located in the prevailing wind direction upstream, it may become too risky to erect a wind turbine there.

Figure 2.21 shows a photo taken by the author during a site visit to an existing but problematic wind farm in China. The wind turbine that is obviously situated in the shadow of the peak is indicated by the rectangle in Figure 2.21. This turbine suffers from the compounding negative effects of reduced wind speed and increased turbulence intensity. Fortunately, the shape of the peak is rather gentle and flow separations may be gentle and as an obstacle its shelter effects may not be as significant. Nevertheless, building wind turbines in such locations should not be encouraged without further investigation. This is where most wind flow models fail to capture the true wind conditions and care should be taken during a micro-siting site visit. For peace of mind, short-term wind monitoring by lidar or sodar is strongly recommended in order to verify whether the turbulence intensity is within the design limits.

In conclusion, obstacle issues may be difficult to resolve by software models but it is an indispensable reference for micro-siting during a site survey. Also, the obstacle effects are usually considered from the perspective of avoiding harmful turbulence, while the calculation of wind resource and energy production need to use specific software and be carried out at a desk.

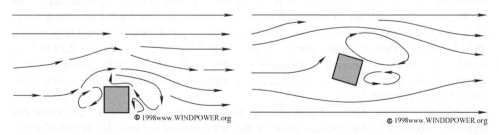

Figure 2.20   Side view (left) and top view (right) of wind flow around an obstacle, indicating more pronounced turbulent airflow downstream [20] *Source*: Danish Wind Industry Association

Figure 2.21 Example of an obstacle-like peak. The wind turbine in the rectangle may suffer from severe turbulence generated by the peak in the front

2.4 Summary

Three types of topographical effects are considered in parallel for wind resource assessment and micro-siting: roughness and surface properties of the terrain; orography and height variations of the terrain; and obstacles and protruding objects on the surface. Their total effects on wind can be regarded as superposition on top of each other for the sake of simplification.

Roughness exerts friction forces on the wind, therefore creating a logarithmic wind profile with wind speed approaching zero near the surface. This model is used to extrapolate the wind speed from one height to another. Normally the roughness lengths of the surface are predefined based on the rule of thumb given in Table 2.1. Wind speed measurements of at least two heights from the same met mast are then used to verify the predefined roughness lengths through cross-interpolation and adjustments should be made until the extrapolated wind speed meets the measured one at the same height within an acceptable margin.

The wind usually accelerates on the crest of a hill, that is the speed-up effect. Flow separations are prone to occur when the slope of the hill is greater than 22° upwind and 17° on the lee, creating complicated nonlinear flow patterns where linearised wind flow models typically exaggerate the speed-up on a hill top. For the purpose of verifying wind flow models and assessing their uncertainties, terrain can be put into four categories (Figures 2.13 to 2.16) based on its complexity level. The Ruggedness Index can be a rudimentary quantification of terrain complexity and is used by WAsP as an attempt to improve the modelling results in the case of flow separations.

Obstacles reduce wind speed and increase turbulence intensity downstream. Obstacles should be avoided with the best endeavour for the siting of wind turbines. Orographic features such as prominent peaks can be treated as obstacles if they are located in the vicinity of the interest site. A two-times-twenty principle (or a three-times-fifty principle for more conservative considerations) is a handy rule of thumb to identify the shelter zone of an obstacle.

References

[1] Troen, I., Petersen, E.L. (1989) *European Wind Atlas*, Risø National Laboratory, Denmark.

[2] Lettau, H. (1969) Note on Aerodynamics Roughness – Parameter Estimation on the Basis of Roughness – Element Description. *Journal of Applied Meterology*, **8**, 828–832.

[3] Kondo, J., Yamazawa, H. (1986) Aerodynamics Roughness over an Inhomogeneous Ground Surface. *Boundary-Layer Meteorology*, **35**, 331–348.

[4] Sempreviva, A.N., Larson, S.E., Mortensen, N.G., Troen, I. (1990) Response of Neutral Boundary Layers to Changes of Roughness. *Boundary-Layer Meteorology*, **50**, 205–225.

[5] Benson, J. (2005) Boundary-Layer Response to a Change in Surface, A Dissertation submitted in Partial Fulfilment of the Requirement for the Degree of MSc Applied Meteorology, The University of Reading, Reading.

[6] Sedefian, L. (1980) On the Vertical Extrapolation of Mean Wind Power Density. *J. Appl. Meteorology*, **19**, 488–493.

[7] Emeis, S. (2013) *Wind Energy Meteorology, Atmospheric Physics for Wind Power Generation*. Springer.

[8] Emeis, S. (2001) Vertical Variation of Frequency Distribution of Wind Speed in and above the Surface Layer Observed by SODAR. *Meteo. Z.*, **10**, 141–149.

[9] Taylor, P.A., Teunussen, H.W. (1987) The Askervein Hill Project: Overview and Background Data. *Boundary-Layer Meteorology*, **1–2** (39), 15–39.

[10] Taylor, P.A., Walmsley, J.L. (1996) Boundary-Layer Flow Over Topography: Impacts of the Askervien Study. *Boundary-Layer Meteorology*, **78**, 291–320.

[11] Kim, H.G, Patel, V.C., Lee, C.M. (2000) Numerical Simulation of Wind Flow over Hilly Terrian. *Journal of Wind Engineering and Industrial Aerodynamics*, **87**, 45–60.

[12] Teunissen, H.W., Shokr, M.E., Bowen, A.J., Green, W.C.J. (1987) Askervein Hill Project: Wind-Tunnel Simulations at Three Length Scales. *Boundary-Layer Meteorology*, **40**, 1229.

[13] Jensen, N.O., Petersen, E.L., Troen, I. (1984) *Extrapolation of Mean Wind Statistics with Special Regards to Wind Energy Applications*, World Meteorology Organization, p.85.

[14] Wood, N. (1995) The Onset of Flow Separation in Neutral, Turbulent Flow over Hills. *Boundary-Layer Meteorology*, **76**, 137–164.

[15] Loureiro, J.B.R., Silva, F.A.P. (2009) Note on a Parametric Relation for Separating Flow over a Rough Hill. *Boudary-Layer Meteorology*, **131**, 309–318.

[16] Martensen, N.G., Bowen, A.J., Antoniou, I. (2006) *Improving WAsP Predictions in (Too) Complex Terrain*, Riso National Laboratory, Denmark.

[17] Miller, C.A., Davenport, A.G. (1998) Guidelines for the calculation of wind speed-ups in complex terrain. *Journal of Wind Engineering and Industrial Aerodynamics*, **74–76**, 189–197.

[18] Loureiro, J.B.R., Monteiro, A.S., Pinho, F.T., Silva, F.A.P. (2008) Water Tank Studies of Separating Flow over Rough Hills. *Boundary-Layer Meteorology*, **129**, 289–308.

[19] Bowen, A.J., Mortensen, N.G. (1996) Exploring the limits of WAsP: The Wind Atlas Analysis and Application Program. *Proceedings of the 1996 European Union Wind Energy Conference*, Sweden, pp. 584–587.

[20] Wind Obstacles, http://www.windpowerwiki.dk/index.php?title=Wind_obstacles, Danish Wind Industry Association (last accessed on 10 December 2013).

3

Numerical Wind Flow Modelling

While analytical models introduced in Chapter 2 are very handy, we will still need more sophisticated wind flow models to simulate winds in order to produce more accurate and quantitative results that are commercially acceptable. Computers become of the upmost importance in this inevitable approach and the models can be scarily complicated in terms of mathematics and physics. Numerous reports and monographs can be found in the regimes of meteorology, computational fluid dynamics (CFD), air flow physics, etc., and it is often puzzling for people outside academics. This chapter, however, is not intended to crack all the wind flow models in great detail, but is rather to consolidate and suffice the fundamentals for technical professionals working in wind resource assessment.

It is generally not practical to measure wind conditions at every potential wind turbine location and at the height of the exact hub height of wind turbines, and in most cases only one or two met masts are erected to represent wind conditions of the entire wind park area at certain heights. Therefore, it is essential to deduce wind conditions from one known location, that is the met mast, to the others where future wind turbines are likely to be constructed, and is therefore for height extrapolation. The purpose of wind flow modelling is to estimate wind resource at each proposed and potential wind turbine location in order to predict the overall energy production of the entire wind park based on one or more meteorological towers.

There is a wide variety of methodologies with diverse characteristics to model wind flows, ranging from simple linear solvers to direct numerical solutions. None of them is perfect however or can claim absolute supremacy over the others. The acknowledgement of the inherent uncertainties of the model in use is critical. It is worth reiterating the stochastic nature of wind, which means that uncertainty is imbedded in each and every procedure of wind resource assessment (see Chapter 7).

Published work in the field of wind flow modelling over topography varies extensively in terms of the choice of topography (two or three dimensional; real or idealised), working variables, turbulence models, coordinate geometry, time dependent or steady state, compressible or incompressible, hydrostatic or nonhydrostatic, and the inclusion of effects such as stability, rotation, latent heat transport and surface fluxes, besides a whole variety of numerical schemes for discretising and solving the governing equations. It is of course impossible to explicitly include all within the scope of this book. What is presented in this chapter is the part that the author finds most helpful with understanding day-to-day engineering problems.

Wind Resource Assessment and Micro-siting, Science and Engineering, First Edition. Matthew Huaiquan Zhang.
© 2015 John Wiley & Sons, Ltd. Published 2015 by John Wiley & Sons, Ltd.

Apart from simulating wind flow variations due to topographical features, flow interactions between neighbouring wind turbines should also be conducted by wind flow modelling. It is usually treated as a separate issue and is thus introduced in Chapter 4.

3.1 Modelling Concept Review

In order to understand different wind flow models, we should have a basic grasp of flow concepts, governing equations and theories. The profound content can be found in many textbooks, such as *Computational Fluid Dynamics* by John (1995) [1], *Fluid Mechanics* by Falkovich (2011) [2] and *An Introduction of Boundary Layer Meteorology* by Stull (1988) [3].

3.1.1 Wind Flow Concepts

Wind flow modelling has to make many assumptions on the flow in order to enable the calculation. Let us look at some of the flow concepts and jargons that commonly appear in those assumptions.

3.1.1.1 Inviscid Flow and Viscous Flow

Viscosity in fluid dynamics is a term used to describe fiction in the fluid in motion. The Reynolds number (*Re*) is defined as the ratio of inertial forces to viscous forces and consequently quantifies the relative importance of these two types of forces for given flow conditions [2]. It can be used to evaluate whether viscous or inviscid equations are appropriate to the problem.

At very low Reynolds numbers ($Re \ll 1$), the inertial forces can be neglected compared to the viscous forces and such flow is called Stokes flow. In contrast, high Reynolds numbers mean that the inertial forces are more dominant than the viscous (friction) forces. Therefore we may assume the flow to be inviscid when the viscous forces are approximately negligible compared to the friction forces.

The Reynolds number of wind flow for energy extraction purposes is very high, easily exceeding the order of millions [4]. This is why wind is usually treated as an inviscid flow in wind resource assessment applications. The very thin (a few millimetres) boundary layer near the ground where viscosity becomes significant is simply too shallow to make a difference.

3.1.1.2 Compressible Flow and Incompressible Flow

All fluids are compressible to some extent; that is changes in pressure or temperature will result in changes in density. However, in many situations the changes in pressure and temperature are sufficiently small that the changes in density are negligible. In this case the flow can be modelled as an incompressible flow. Otherwise the more general compressible flow equations must be used [1].

Mathematically, incompressibility is expressed by saying that the density ρ of a fluid parcel does not change as it moves in the flow field. This additional constraint simplifies the governing equations, especially in the case when the fluid has a uniform density.

Nearly all engineering applications of wind resource assessment assume the wind flow to be incompressible, an acceptable assumption in the atmospheric boundary layer. It is obviously not true in real case scenarios but one of many compromises we have to make to model wind flows.

3.1.1.3 Steady Flow and Unsteady Flow

Steady-state flow refers to the condition where the fluid properties (velocity, pressure, temperature, etc.) at a point in the system do not change over time. Otherwise, flow is called unsteady (also called transient [2]).

Steady or unsteady is a relative term depending on the chosen frame of reference. According to Pope (2000) [5], an unsteady turbulent flow can be statistically stationary if all statistics are invariant under a shift in time. This roughly means that all statistical properties are constant in time. More often than not, the mean flow field is the object of interest in wind resource assessment, and this is also constant in a statistically stationary flow.

Steady flows are often more tractable than otherwise similar unsteady flows. The governing equations of a steady problem have one dimension fewer (time) than the governing equations of the same problem of an unsteady flow.

With these benefits, it is not hard to imagine steady flow to be one of the common implications for engineering applications of wind resource assessment.

3.1.1.4 Laminar flow and turbulent flow

Turbulence is flow characterised by recirculation, eddies and apparent randomness. Flow in which turbulence is not exhibited is called laminar. Laminar flow is smooth with adjacent layers of fluid sliding past each other without intermingling. Turbulent flow is a much more common form of fluid flow. The wind flow in the boundary layer of the atmosphere is a good example of turbulent flow.

Mathematically, turbulent flow is often represented via a Reynolds decomposition, in which the flow is broken down into the sum of an average component and a perturbation component.

The Reynolds number can also be used to indicate whether the flow is prone to turbulent or laminar flow; the larger the Reynolds number, the more turbulent the flow. Winds are certainly in the region of very turbulent flows.

3.1.2 Governing Equations

The governing equations of flow dynamics are as applicable to atmospheric flows as they are to any laboratory experiments. Wind flow modelling normally does not try to solve all of the governing equations simultaneously while undertaking considerable levels of simplification. Simpler models may be more robust and undeniably require less resources and expertise to run.

Each meteorological variable (e.g. wind speed, temperature, pressure) can be written as a mean plus a superimposed fluctuating value (turbulence) [3], that is

$$X = \overline{X} + X' \tag{3.1}$$

where $\overline{X}$ denotes the mean term and X' the fluctuation term. The best thing about this separation is that $\overline{X}$ and X' can be modelled and studied separately.

3.1.2.1   Mean Flow

The mean flow equations involved are the ideal gas law and the laws of conservation (including conservation of mass, momentum, energy and other scalar quantities), which can be manipulated and constructed in various forms.

Conservation of mass is also known as the mass continuity principle. The rate of change of fluid mass inside a control volume must be equal to the net rate of fluid flow into the volume. Physically, this statement requires that mass is neither created nor destroyed in the control volume. Conservation of momentum applies Newton's second law of motion to the control volume, requiring that any change in momentum of the air within a control volume must be due to the net flow of air into the volume and the action of external forces on the air within the volume. Conservation of energy means that although energy can be converted from one form to another, the total energy in a given closed system remains constant, which is the application of the first law of thermodynamics to a control volume.

Since wind speed is the prime pursuit in wind engineering, solving momentum equations becomes the centre of focus for any wind flow models. The momentum equations for Newtonian fluids are the Navier–Stokes equations, which are a nonlinear set of differential equations that describes the flow of a fluid whose stress depends linearly on velocity gradients and pressure. The fundamental basis of almost all CFD problems is the Navier–Stokes equations.

The standard Navier–Stokes equations were originally derived for laminar flow. Reynolds (1895) [6] suggested an approach to split them into a mean component and a fluctuating value (see the following Equations (3.2) to (3.5)) in order to represent the turbulent variations. This is a more accurate representation of the real fluid flow, but it gives rise to additional terms in the equations, that is the Reynolds stress or otherwise called turbulent flux. For steady and incompressible flow, those equations become

$$\frac{\partial U}{\partial x} + \frac{\partial V}{\partial y} + \frac{\partial W}{\partial z} = 0 \tag{3.2}$$

$$U\frac{\partial U}{\partial x} + V\frac{\partial V}{\partial y} + W\frac{\partial W}{\partial z} = -\frac{1}{\rho}\frac{\partial P}{\partial x} + \frac{\partial \overline{u^2}}{\partial x} + \frac{\partial \overline{uv}}{\partial y} + \frac{\partial \overline{uw}}{\partial z} \tag{3.3}$$

$$U\frac{\partial U}{\partial x} + V\frac{\partial V}{\partial y} + W\frac{\partial W}{\partial z} = -\frac{1}{\rho}\frac{\partial P}{\partial y} + \frac{\partial \overline{uv}}{\partial x} + \frac{\partial \overline{v^2}}{\partial y} + \frac{\partial \overline{vw}}{\partial z} \tag{3.4}$$

$$U\frac{\partial U}{\partial x} + V\frac{\partial V}{\partial y} + W\frac{\partial W}{\partial z} = -\frac{1}{\rho}\frac{\partial P}{\partial z} + \frac{\partial \overline{uw}}{\partial x} + \frac{\partial \overline{vw}}{\partial y} + \frac{\partial \overline{w^2}}{\partial z} \tag{3.5}$$

where U, V and W are the components of the mean wind vector in three directions x, y and z respectively; u, v and w the turbulence fluctuation components; P the mean pressure value; and ρ the air density. Equation (3.2) is the continuity equation and Equations (3.3) to (3.5) are the conservation of momentum.

Here $\overline{uv}$, $\overline{uw}$ and $\overline{vw}$ are the Reynolds *shear* stresses and $\overline{u^2}$, $\overline{v^2}$ and $\overline{w^2}$ are the Reynolds *normal* stresses. The appearance of these unknown Reynolds stress terms creates a problem of closure. The Navier–Stokes equations are no longer solvable directly as the 10 unknown terms are present for the same number (four) of equations of motion. Turbulence models must be introduced for the solution of the flow problem, also known as turbulence closure technics. The Navier–Stoke equations are then renamed the Reynolds averaged Navier–Stokes (RANS) equations.

3.1.2.2   Turbulence Models

A turbulence model is a mean of specifying the Reynolds stresses and turbulent fluxes, hence closing the mean flow equations. A host of turbulence closure models has been developed over the years and numerous monographs dedicated to this subject that readers can refer to (e.g. Pope, 2000 [5]). Without getting lost in the maze, there are a few basic and most recalled concepts and ideas that are worth knowing for a better understanding of wind flow modelling.

Turbulence is initially generated by instabilities in the flow caused by the mean velocity gradient. These eddies in turn breed new instabilities and hence smaller eddies and then even smaller eddies until the ultimate dissipation of turbulence energy by viscosity. This turbulence creation process is called the turbulence energy cascade. Viscosity only becomes important when turbulence eddies are sufficiently small and start dissipating into heat, that is vibration of air molecules. Turbulence models for the purpose of wind resource assessment usually deal with relatively large turbulence eddies and hence viscosity of the air can be ignored in most cases.

Turbulence can be transported or carried with the flow, but at high Reynolds numbers (e.g. the air) turbulence transported a far greater amount of momentum than viscous forces throughout most of the flow, which is another good reason for neglecting viscosity of the air.

There is a general hierarchy in turbulence models. In principle, the higher in the hierarchy the more general or accurate the turbulence model becomes, but one step higher demands so much more computational effort and resources that it may eventually outweigh any potential benefit. The hierarchy can be generalised from the lowest to the highest as the mixing-length model (first-order closure), eddy viscosity models (EVMs), Reynolds stress transport models (RSTMs, second-order closure), large eddy simulation (LES, second- and higher-order closures) and direct numerical simulation (DNS).

The mixing-length model was introduced by Prandtl (1925) [7] and represents the distance travelled by a small discrete volume of fluid before it becomes part of the mean flow. Mixing length is supposed to be unique and characterises the local turbulence intensity and the size of turbulence eddy and can be a function of position or velocity. Mixing-length models work well in near-equilibrium boundary layers, but are difficult to generalise to more complex flows, as they fail to account for transport of turbulence quantities such as diffusive and convective transport, which are important in the case of flow separation and recirculation. The mixing-length model is essentially a zero equation EVM as it solves zero turbulence transport equation.

The κ–ε model is probably the commonest turbulence model in use today. It is a two-equation EVM model. Here, κ represents the turbulence kinetic energy ($\kappa = 0.5(\overline{u^2} + \overline{v^2} + \overline{w^2})$) derived from the Navier–Stokes equations and ε, the rate of dissipation of turbulence kinetic energy, is on the other hand heavily modelled [5]. The κ–ε model is not a single model but a class of slightly different schemes arising from modifications of viscous effects through the Reynolds number. EVMs are popular because they are simple to code and supported theoretically in some simple but common types of flow, but they fail to represent turbulence physics. The EVMs can also be easily 'tweaked' for particular types of flow.

The standard $\kappa-\varepsilon$ model does not generally perform well in wind engineering because it only works properly when normal Reynolds stresses are not important (homogenous isotropic flow), which is not valid in wind engineering, especially for complex terrain. By definition, the turbulence kinetic energy κ that the $\kappa-\varepsilon$ model predicts is isotropic. It is undoubtedly not true in wind engineering as the wind flow is definitely anisotropic [8]. The $\kappa-\varepsilon$ model also tends to overestimate κ. Many modifications to the standard $\kappa-\varepsilon$ model have been proposed and tested to combat these and other issues, such as the $\kappa-\varepsilon$ RNG model proposed by Yakhot and Orzag (1986) [9], the low Reynolds number $\kappa-\varepsilon$ model where viscous stresses replace the Reynolds stresses close to the wall (Patel *et al.*, 1985 [10]; Apsley, 1995 [11]; Rahman and Siikonen, 2002 [12]), to name just a few.

The RSTMs, or second-order closure or differential stress models, solve individual transport equations for all stresses rather than just the turbulence kinetic energy κ. Both the advection term (turbulence carried by the mean flow) and the production term (creation of turbulence by the mean flow) are exact and not modelled; thus the RSTMs should take better account of turbulence physics than EVMs. However, six turbulence transport equations make the models very complex and computationally expensive. RSTMs also tend to be numerically unstable because only small molecular viscosity contributes to any sort of gradient diffusion.

The LES technique is gradually becoming a common tool in research studies of atmospheric boundary layers [13]. LES applies low-pass filtering to the Navier–Stokes equations to eliminate small scales of the solution, reducing the computational cost of the simulation. The governing equations are thus transformed and the solution is a filtered velocity field, of which the 'small' length and time scales to eliminate are selected according to turbulence theory and available computational resources [5, 14]. One of the problems associating LESs is that the turbulence kinetic energy of scales smaller than the filter is not resolved. The cost of refining the filter scale is extremely high.

Detached eddy simulation (DES) is therefore proposed. In wind engineering, this means that for simple flow situations without separations, the simulation reduces to a simpler model and LES attributes are only introduced as the flow becomes more complicated. DES is still computationally demanding but is a promising method for strongly separating flows [15].

DNS is not a turbulence model *per se*. It is a solution of the complete time-dependent Navier–Stokes equations without a turbulence model. DNS is prohibitively expensive in most majority cases and thus of little interest for wind engineering today.

3.1.2.3   Additional Atmospheric Equations

Coriolis Force

The rotation of the earth has consequences on atmospheric flows resulting in forces known as the Coriolis forces. They add an extra term to the governing flow equations. Coriolis forces are normally neglected in the boundary layer as they only become significant in the outer layer or at the top of the boundary layer.

Hydrostatic Equation

Temperature and pressure of the atmosphere varies with height, which is expressed by the hydrostatic equation. It assumes that the atmosphere is free from vertical acceleration and the mean flow.

Heat Balance

As the wind flow model becomes more meteorological and less fluid dynamic, other equations that are related to the heat balance of the atmosphere have to be included as well, such as latent heat arising from vaporisation and condensation of water, convections that vertically transport heat and mass, solar and infrared radiation or heat balance of the surface, etc.

Surface Friction

To simulate winds at the lower boundary layer, surface friction or roughness has to be carefully modelled as well.

Chapters 9 and 10 will elaborate on this topic.

3.1.3 Meshing the Computational Domain

Meshing is a process that divides the computational domain into a large number of mesh elements (cells) that tessellate to fit to the surface in the domain, so that the governing flow equations can be integrated over each element to form a discrete equation at its nodes. There are generally two grid types, structured and unstructured. Structured meshing defines consistent labelling of the elements in rows, columns and levels where the equations are solved for each element in relation to its surrounding elements. Unstructured meshing does not annotate meshing elements in the same way and the method of solution differs. Unstructured meshing is more complex but more adaptable to complex boundary conditions.

For wind engineering, the arbitrary surface shape usually requires a meshing technique that allows the meshing elements to fit to the shape of the orography and the meshes should be much more refined vertically closer to the surface (see Figure 3.1) because we only care about

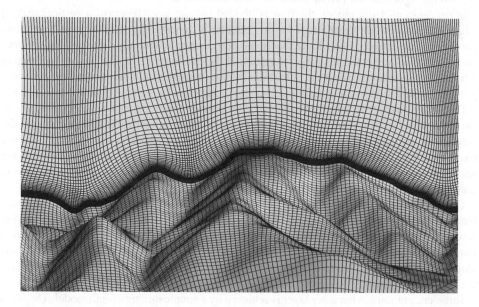

Figure 3.1   Example of a structured mesh (WAsP CFD) [16] *Source*: Risø National Laboratory

the wind field at the lower level, that is the wind turbine hub height; it is also a common practice to use finer grids around the interest sites horizontally in order to produce more accurate simulations.

Finer grids usually give a more accurate simulation, but at the same time it demands exponentially more computational resources. Therefore, it is usually a trade-off between the mesh resolution and the computational resources available. The model may also become unstable under grid sizes that are too refined. Meshing is actually the most difficult part for constructing the model and requires substantial expert knowledge.

3.2   Linearised Numerical Flow Models

The inherent nonlinearity of the Navier–Stokes equations makes them really difficult to manage and exact analytical solutions are rarely available. It is therefore not unreasonable to try our best to linearise the equations of motion. One of the most intriguing properties of linearised theory is that it allows superposition, enabling the perturbation for randomly shaped topography to be generated from the integral sum of individual Fourier modes.

3.2.1   Jackson–Hunt Model

Turbulence is one of the most important features for wind flow in the boundary layer of the atmosphere. Jackson and Hunt (1975) [17] proposed a linearised model of Navier–Stokes equations, describing two-dimensional turbulent air flow over low hills in the boundary layer. It remains the foundation of most linearised flow models, even though it has undergone considerable refinement. WAsP, the pioneering and most widely used wind resource assessment application, is also developed upon the Jackson–Hunt model.

In the Jackson–Hunt model, the flow is divided into an inner layer (of a vertical scale l) where the perturbation shear stress (terrain effects) is dynamically significant and turbulent prevails and an inviscid outer layer where the perturbation essentially vanishes. Streamlines are forced up and displaced by height variation of the terrain (terrain perturbation) (see Figure 2.6) in the lower layer. Together with atmospheric thermal stratifications, it then generates pressure perturbation in the outer layer, which in turn feeds back to the flow at a lower level.

The terrain shape (perturbation) is analysed in terms of Fourier components. The equations are thus solved in Fourier space. The Fourier transform is numerically inverted to give the solution in real space. The lower-level flow is modified by turbulent transport and surface roughness. The equations of motion are linearised in each layer by writing velocity as the sum of the upstream velocity and a small perturbation, and the zeroth- and first-order terms in matched Fourier expansions respectively.

The model requires the horizontal scale of the perturbation to be much less than the distance required for the boundary layer to change noticeably (small perturbation assumption). In wind resource assessment terms, this also means that the project area has to be under the same wind climate; the maximum height of the hill is much smaller than the length scale of the hill and the slope of the hill is assumed to be small everywhere [17], that is a simple terrain (see Figures 2.13 and 2.14).

Turbulence closure in the inner layer is achieved by a simple mixing-length model. There is a mismatch between the inner layer and the outer layer in terms of the longitudinal approach-flow velocity. Two alternatives have been proposed to improve on this and are adopted individually by FLOWSTAR and MS3DJH codes.

The FLOWSTAR model has incorporated an approach proposed by Hunt *et al.* (1988) [18, 19], where the inviscid outer layer is subdivided into an upper layer and a middle transient layer. The effect of approach-flow velocity shear is manifested in the middle layer although neglected in the upper layer [20].

Mason and Sykes (1979) [21] extended the Jackson–Hunt model to three dimensions and developed the MS3DJH (the Mason and Sykes three-dimensional extension of Jackson–Hunt theory) code by adopting the alternative approach. In this approach, the concept of a wavenumber scaling technique is introduced to account for the large range of horizontal scales of real terrain rather than just one in the Jackson–Hunt model. The MS3DJH model is adopted by WAsP [22].

Linearised perturbations induced by surface roughness are added linearly to those generated by height variations in both FLOWSTAR and MS3DJH/3R codes.

The mixing-length model is rudimentary when it comes to describing turbulence. MS-MICRO, the personal computer implementation of MSFD (mixed spectral finite-difference) model [23] therefore combined the Jackson–Hunt spectral approach in the horizontal with a finite-difference solution (linearised $\kappa-\varepsilon$ model) in the vertical. A more general turbulence model was achieved without large computational overheads. The MS-Micro/3 model is adopted by WindFarm [24], a well-known wind resource assessment application. A nonlinear extension of the MSFD model (NLMSFD) has been proposed by Xu and Tayler (1992) [25].

Inspired by the linear theory of Jackson and Hunt (1975) [17] and the MSFD family of models [23], the RAMSIM, a new micro-scale, linearized flow model, has been proposed by Corbett in his PhD thesis [26] to calculate the flow field over terrain comprising steep slopes without compromising too much on computational costs. The RAMSIM model is governed by the RANS for incompressible flow in the neutrally stratified atmospheric boundary layer and the $E-\varepsilon$ closure equations expressed in general curvilinear coordinates. These equations are linearised by a perturbation expansion, of which only the zero- and first-order terms are retained. The zero-order equations simply describe logarithmic wind flow over a flat terrain of roughness z_0, while the first-order equations represent the spatial correction due to the presence of orography. A terrain-following coordinate system is based on a simple analytical expression, allowing rapid computation. According to the test results in the PhD thesis, RAMSIM can, to a certain extent, predict the occurrence of recirculation, thus setting itself apart from previous linear models.

3.2.2 WAsP Model: The Principle

The Wind Atlas Analysis and Application Program (WAsP) developed at the Risø National Laboratory in Denmark in 1987 is a spectral model belonging to the family of the Jackson–Hunt theory, or the MS3DJH family to be more exact [27]. The WAsP model is often referred to as the European Wind Atlas method [28]. WAsP is very successful and can be treated as the industrial standard for wind resource assessment purposes. Some renowned commercially available products, such as WindPRO and WindFarmer, make direct use of the WAsP results. The main principles are also implemented on a rectangular grid basis in the LINCOM model [29], which is implemented in WAsP-Engineering, the sister product of WAsP. Therefore, the WAsP model deserves the spotlight of wind resource assessment. The diagnostic property of the model is also of great help for our day-to-day engineering work. This section will elaborate on the ideas of WAsP.

Wind near the surface is modified by the terrain's micro-scale (local) features (terrain perturbations). For two locations not too far from each other, these micro-scale flow features are assumed to be superimposed on their common wind climate of larger scales corresponding to the outer layer where perturbations are negligible. This is the fundamental principle of WAsP through which a link between wind speeds at two locations within the same wind climate is established. The common regional wind climate is obtained by peeling off these micro-scale modifications on the wind data measured at the reference site. The wind climate of a specific site is then calculated by adding the specific micro-scale modifications back to the common regional wind climate. Figure 3.2 describes this calculation process. The perturbations of the terrain are modelled by three parallel models: the obstacle sheltering model, the roughness model and the orographic model (speed-up effects), as introduced in Chapter 2.

Mathematically, the process of deducing the common regional wind climate is

$$W_A = W_{reg} - OBS_A - ROU_A - ORO_A \qquad (3.6)$$

where W_{reg} represents the common regional wind climate, W_A the micro-scale wind climate of location A and ORO_A, ROU_A and OBS_A represent the effects of orography, roughness and obstacle respectively.

Similarly, the wind climate of location B within the same regional wind climate is

$$W_B = W_{reg} + OBS_B + ROU_B + ORO_B \qquad (3.7)$$

$$W_B - W_A = (OBS_B - OBS_A) + (ROU_B - ROU_A) + (ORO_B - ORO_A) \qquad (3.8)$$

The surface roughness model in WAsP is based on Charnock (1955) (see Section 2.1.4). As for the Jackson–Hunt model, WAsP is a spectral model and solves the linearised Navier–Stokes equations with only first-order velocity perturbation induced by the terrain taken into account in a Fourier space. The Fourier space implemented in WAsP is the so-called BZ (Bessel expansion on a zooming grid) flow model [27, 30].

The BZ model is specially designed for wind turbine micro-siting with many unique features [31]; it is a high-resolution zooming grid (see Figure 3.3), which calculates the potential flow perturbation profile at the centre; it integrates the surface roughness model into the spectral; the inner layer velocity field is simulated using a balance condition between surface stress, advection and the pressure gradient; and large scale flow is forced to the upper layer by fixing the boundary layer thickness at approximately 1 km.

The model uses 100 radial stations and 72 radii (5° spacing) to create the Fourier space where the Fourier transformation of velocity perturbations is performed. Unlike other models, which only mesh once for the computational domain, the BZ model meshes individually for each site.

The nature of the zooming grid shows that the resolution of the grid is the highest at the centre and decreases outwards. This means that the resolution of the height contour lines in use should be of a higher resolution closer to the centre of interest in order to capture more orographic information, but a high contour resolution is not necessary far off the centre. Mapping detailed height contour lines of the entire project area can be expensive and time-consuming in a complex terrain. It would be wise to only map adequate contours of the potential ridges

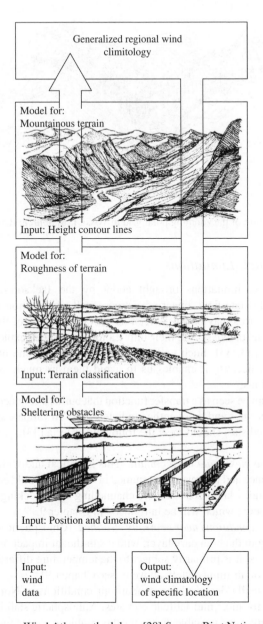

Figure 3.2   The European Wind Atlas methodology [28] *Source*: Risø National Laboratory

(1 ~ 2 km radius) and then combine it with a coarser satellite contour to cover the entire region as far as the BZ model is concerned.

The borders of the contour map are recommended to be at least 5 km distant from the centre of interest sites covering an area of at least $10 \times 10 \, km^2$ with a height contour interval of less than 20 m. If only a digital elevation model is available, the grid size should be less than 50 m, not critical if the terrain within 1 km of the site is digitalised in detail [32].

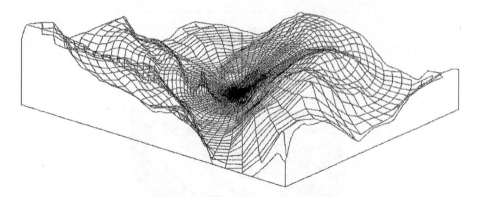

Figure 3.3   Grid structure of the zooming polar grid in WAsP [28] *Source*: Risø National Laboratory

3.2.3   WAsP Model: Limitations

WAsP inherits all the limitations brought about by the Jackson-hunt model, such as neutrally-stable steady-state flow, linear advection, small perturbation and first-order turbulence closure. Great errors may occur if WAsP is operated outside of its comfort zone. Over the year, the limitations of WAsP and their implications have been explored quite thoroughly and well understood [33, 34]. Such understanding has become one of the most important advantages of WAsP, because it enables us to avoid them as much as possible or at least analyse the possible bias.

The WAsP model can be seen as a transfer function that builds a connection between the wind speeds at the reference sites and those at the predicted sites. To validate such a connection, the reference and predicted sites should meet the following criteria [33]:

1. *Subject to the same mesoscale wind climate*. This is critical and without it the connection between the reference and interest sites assumed in the model would collapse. A meso-scale wind climate includes land–sea breezes, channelling and blocking of large mountains, downslope winds, etc., which will be introduced in Chapter 9.
2. *Prevailing weather conditions near neutrally stratified*. Atmospheric stability characterises the level of mixing in the boundary layer, with a significant impact on the mean wind flow and speed-up effects. It is primarily subject to the temperature difference between the surface and the layer of air immediately above it (see Chapter 9).

   WAsP was originally designed for standard air conditions in Northern Europe, which are representative for most mid-latitude climates. Atmospheric stability varies constantly: prone to stable at night and unstable during the day; stable in winter and unstable in summer. The atmospheric stability is taken as an annual mean, that is the prevailing stability, in WAsP. The heat flux parameter in WAsP should be adjusted with caution if nonstandard atmospheric stability is confirmed.
3. *Gentle surrounding terrain to ensure mostly attached flow*. The linearised flow model implemented in WAsP assumes that the wind flow is attached to the surface without separation. It works well under gentle terrain conditions of slopes smaller than 0.3 (17°), but when slopes are greater than 0.3, the possibility of flow separation becomes tangible and the speed-up effect no longer follows a linear trend, as presented in Figure 2.9. Here the flow is attached to the virtual surface with slopes smaller than those of the real terrain

surface (see Figure 2.12), indicating an overestimation of the speed-up effect. The error induced by the surrounding terrain, being outside the recommended operational envelope, can be significant and overturning.

4. *Reliable reference wind data*. The accuracy of modelling wind speeds using WAsP relies heavily on the quality of the wind data, more so than when using more complete flow models. The reliability of wind data is largely reflected by (see Chapter 8):

   - Representativeness of the reference wind data;
   - Installation specification and maintenance of the meteorological mast;
   - Recovery of the wind data and missing data, which should be evenly distributed in the entire time series;
   - At least one whole year of wind data covering seasonal variations of the wind climate.

5. *Overestimating sheltering effect of three-dimensional obstacles*. A three-dimensional obstacle (e.g. a house) can be regarded as a change in elevation of the terrain to some extent. There is a speed-up zone on the top of and downstream from the obstacle, which is not simulated by the obstacle model implemented in WAsP. The two-dimensional obstacle sheltering model also has a tendency to overestimate the wind speed reduction (see Section 2.3).

6. *Model independence hypothesis*. The WAsP model consists of three independent models, that is the obstacle sheltering effect model, the roughness change model and the terrain speed-up model. These three linearised models are assumed to be interindependent and therefore allowing direct calculation of the sum arithmetically (see Equations (3.6) to (3.8).

   In reality, these geographic effects on the wind are usually not that clear-cut. For instance, a tree can be treated as a roughness element but may also become an obstacle in some cases. Due to the complexity of the micro mechanisms of the wind flow, studying them separately, especially for engineering purposes, is inevitable.

7. *Logarithmic vertical wind profile*. The logarithmic wind profile (see Section 2.1.2) only characterises the vertical wind profile in the surface layer of the boundary layer and becomes invalidated in the Ekman layer above it. It has not been an issue until recent years because wind turbines were previously much smaller in size and lower in hub height, which means that they would operate mostly within the surface layer. However, wind turbines nowadays can easily exceed a hub height of 80 m with a rotor diameter of over 100 m. At least part of the wind turbine's sweeping area is not submerged by the surface layer and is exposed to a much more complex vertical profile (see Chapter 10). Events like low-level jets with protruding wind profiles are very likely to become a norm for large wind turbines. This is the reason why the study of boundary layer meteorology is attracting increasingly more attention.

   Attention should therefore be paid to the robustness of vertical extrapolation of wind speeds using WAsP for large size wind turbines.

Example 3.1

A house with an obvious sheltering effect on the reference wind data has been ignored when modelling the wind flow with WAsP. The project area is within the operational envelope of WAsP. Does the predicted wind speed at the wind turbine hub height tend to be overestimated or underestimated?

Solution

Ignoring the house's shelter effect on the reference wind data would be equivalent to ignoring the induced wind speed reduction, which in turn would result in an overestimated regional wind climate (see Equation (3.6)). Therefore the wind speed at the wind turbine hub height is overestimated.

Example 3.2

If the slopes surrounding the reference site are greater than 0.3 while those surrounding the wind turbine location are not, does the energy production of the wind turbine tend to be overestimated or underestimated?

Solution

Slopes of greater than 0.3 indicate the occurrence of flow separations and overestimation of the terrain speed-up effect. Following Equation (3.6), the common regional wind climate is overcompensated and thus underestimated, consequentially underestimating the energy production of the wind turbine.

The examples above have demonstrated the advantage of understanding the principles and limitations of WAsP over other more complicated nonlinear wind flow models. We can easily analyse the accuracy of the results and their tendencies.

Given its limitations, WAsP has been proved to produce good results for the prediction of offshore flow fields [35].

3.2.4   WAsP Model: Improving the Results

In order to improve the accuracy of WAsP predictions in a complex terrain, Mortensen *et al.* (2006) [36] apply statistical corrections based on the terrain's Ruggedness Index (RIX), introduced in Section 2.2.4. RIX is a coarse measure of the extent of flow separation and the extent of the bias of WAsP predictions induced by orography. ΔRIX is defined as the RIX difference between the predicted and reference sites, that is

$$\Delta RIX = RIX_P - RIX_M \tag{3.9}$$

where RIX_P and RIX_M represent the RIX values at the predicted and met mast (reference) sites respectively.

ΔRIX compares the ruggedness of the reference and predicted sites. $\Delta RIX \approx 0$ means that the two sites are equally rugged with a similar level of bias, which indicates that the prediction errors are relatively small. If $\Delta RIX > 0$, WAsP would generally overestimate the wind speeds of the wind turbines sites, or vice versa [34] (also see Example 3.2 above).

Mortensen *et al.* (2006) [36] studied the Joule programme project based in Northern Portugal [37] and mathematically established the relationship between ΔRIX and WAsP wind speed prediction error. The Portugal project provides extensive data for wind resource study in a complex terrain. Five met masts from the Portugal project were used by Mortensen *et al.* (2006) [36] to conduct cross-predictions, resulting in 25 predicted mean wind speeds. The prediction error is written as

$$Err = \ln\left(\frac{U_P}{U_M}\right) \tag{3.10}$$

where U_P is the predicted wind speed and U_M the measured wind speed.

The results are plotted in Figure 3.4. The fitted line in the figure is close to linear and the best-fit line crosses the axes at (0, 0) [36].

Taking α as the slope of the regression line, then Equation (3.6) can be transformed into

$$U_M = \frac{1}{\exp(\alpha\,\Delta RIX)}U_P \tag{3.11}$$

The purely empirical equation above provides the correction factor that could be applied to the biased wind speed predictions in order to obtain the true wind speed at the same height. For the Portugal project analysed by Mortensen *et al.* (2006) [36], the WAsP energy production estimates are improved by 69% (100% corresponds to no error) on average using the correction methodology. For site pairs with ΔRIX larger than 5%, the cross-predictions are improved by over 90% on average.

In order to achieve the regression slope α, there must be at least two met masts in the project area. This emphasises the importance of setting up multiple met masts in a complex terrain so that improvements on the WAsP predictions can be made.

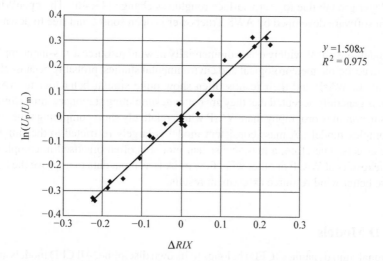

Figure 3.4 WAsP wind speed prediction error as a function of ΔRIX. Square dots are the log errors of the 25 predicted wind speeds and the solid line is the fitted line [36] *Source*: Risø National Laboratory

3.3 Mass-Consistent Models

Another diagnostic approach (reconstructing a steady-state wind field from a set of initial experimental data without time-dependent terms) to model mean wind flow is to construct a mass-consistent wind field from observational data. This is called mass consistent because it solves only one governing equation of motion, the conservation of mass, and therefore does not require large computational resources.

In a mass-consistent model, all variables are defined on a numerical grid with grid increments (Δx, Δy and Δz) in the three dimensions. The three-dimensional wind field is reconstructed using a two-step procedure. Firstly, the available sparse wind-filed data (velocity and direction) are interpolated on these grid points with a weighting factor of some kind as a first guess without considering physical processes or satisfying any conservation equations. Then the interpolated wind field is adjusted to satisfy the incompressible form of mass conservation (constant air density) by minimum possible modifications. By definition, those models cannot describe thermally induced wind systems, such as lans–sea breezes and mountain–valley circulations.

The model is sensitive to the quality of the first guess, which is by and large determined by the interpolation method used and the quality of the wind data. The same data may produce different wind fields by changing only the interpolation method. The effectiveness of these models over complex terrains depends on the availability of properly sited input data. An additional input data location logically improves the first guess. However, optimizing the topographic position of the input data is assumed to have greater importance than a limited increase in the number of input data.

Many mass-consistent wind flow models have been developed since the late 1970s. Examples include NOABL [38, 39], MATHEW [40, 41], MINERVE [42], COMPLEX [43], WINDS [44], WindMap [45] and more. The WindMap/openWind model, like probably all other mass-consistent models, is based on the NOABL model. It takes into account internal boundary layer growth due to sharp surface roughness change [45, 46]. The openWind wind farm design software developed by AWS Truepower is open source and free to download and use [47].

Those models are not as widely used commercially in wind resource assessment applications as they are in academic meteorological and environmental studies, probably because they have similar limits to WAsP and their results often show quite similar behaviour to WAsP [48]. However, it is generally accepted that they are better at simulating stationary three-dimensional wind fields in complex orography than WAsP with a relatively short computing time compared to more complex models. A mass-consistent model is largely restricted to the simulation of orographic effects. Therefore, a mass-consistent model is often practised in complex terrain by using the results of WAsP or other wind flow models as input data to improve the first guess and achieve better wind resource assessment results.

3.4 CFD Models

Computational fluid dynamics (CFD) belongs to its own discipline. All CFD models are essentially RANS equation solvers. They solve more complete forms of RANS equations compared to the Jackson–Hunt model without linearisation and making assumptions like small perturbations, attached flow and first-order closure. Therefore, in theory, they are capable of simulating nonlinear flow phenomena, such as flow separation and recirculation, and are expected to perform better in complex terrain. They also simulate topographical effects as a whole rather than relying on separate modules as WAsP does.

CFD models are computationally much more expensive than WAsP. General-purpose CFD packages, like PHOENICS, CFX, FLUENT, OpenFOAM and STAR CCM+, are applicable for steady or unsteady, one-, two- or three-dimensional turbulent or laminar, compressible or incompressible, viscous or inviscid flows and more, as well as flexible meshing options. These CFD packages required extensive expert knowledge to set up for a specific type of flow situation (in this case wind flow modelling in the atmospheric boundary layer), let alone the cost of running supercomputer clusters. The results are usually heavily dependent on the user's experience, a bad thing for wind resource assessment as it means the results are rarely repeatable by others who also evaluate the same project, which in turn makes the project less bankable.

Many large organizations have established in-house expertise and facilities necessary to develop and run customised CFD modules based on the general-purpose CFD packages. The Vestas in-house CFD toolkit, for example is based on the open-source OpenFOAM code [49]. Visualisation is also one of the key advantages of CFD results for wind turbine micro-siting purposes. Those heavy CFD models are especially good at simulating wind load conditions in terms of turbulence intensity, inflow angle, wind shear and wind direction shift in complex terrain, which is why it is so welcomed by wind turbine manufacturers who have to provide a warranty for their products.

Because CFD models primarily solve the RANS and continuity equations, thermal conditions with respect to atmospheric stability and thermally induced circulations are not resolved, which lead to assumptions like a neutrally stratified atmosphere (which may be tuneable by simple parameters like heat flux or turbulence length scale) and a logarithmic law vertical wind speed profile and prescribed equilibrium inflow conditions, not unlike the Jackson–Hunt and mass-consistent models. Although limited to available computational power, steady-state flow is often assumed as well.

3.4.1   Meteodyn WT and WindSim

In order to enable wind resource assessment engineers with little expert knowledge in the field to operate a CFD model on a workstation, it is very necessary to develop: a robust user-independent mesh solution that is adaptable to all kinds of arbitrarily shaped complex terrain and suitable for wind turbine siting purposes; predefined lighter governing equations (usually incompressible and steady-state flow) and a turbulence closure solution; and a solution to include thermally induced forces in order to describe the flow in the atmospheric boundary layer. Two CFD-based wind farm resource assessment applications, that is Meteodyn WT and WindSim, have made it into the mainstream to some extent in the last decade.

Meteodyn WT, developed by Meteodyn, solves the nonlinear Navier–Stokes momentum equation with the MIGAL solver [50], which implements a 'coupled resolution' that simultaneously updates the wind speed components and the pressure on the whole computational domain using an iterative linear equation solver. Meteodyn WT assumes an incompressible and steady-state flow. It uses a one-equation κ turbulence closure scheme based on Yamada (1983) [51] and Arritt (1987) [52]. The turbulence model allows the solution to take into account the wind turbine wake. Instead of solving the conservation of energy equation to resolve thermally induced effects, thermal stratification of the boundary layer is taken into account by computing an adjusted turbulence length scale at the beginning of the calculation. Ten thermal stability classes are given in the model with neutral atmospheric conditions as the default. Surface roughness is used to determine the initial logarithmic wind speed profile in the surface layer at the computational domain inlet and an Ekman wind speed profile for the remainder of the atmospheric boundary layer. In Meteodyn WT, the turbulence intensity is estimated by the ratio between the square root of the turbulence kinetic energy and the local speed of the flow. A directional speed ratio between every point and the reference site is calculated in post-processing. The wind speed value of each point is then computed by applying those ratios to the reference wind speed.

The WindSim package (developed by the company WindSim AS of Tonsberg, Norway) is based on the PHOENICS code and solves nonlinear transport equations of mass, momentum and energy, providing a more complete solution of motion equations. The simulation begins with a logarithmic wind profile determined by surface roughness and thermal stability as a lateral boundary condition in the assumed atmospheric boundary layer. Above it, a constant wind speed (10 m/s as the default, typical of the geostrophic wind speed at mid latitudes) is considered. The boundary condition at the top of the computational domain can be selected manually as 'fixed wall' (no friction condition) for flat areas or 'fixed pressure' (zero relative pressure), which is better for sites with height variations. The wall with no friction fixes a null value for the vertical component of the velocity while the other velocity component and the pressure are calculated, behaving like a physical wall where the viscous tangential forces are equal to zero. The zero relative pressure boundary condition is an outflow boundary condition where the pressure is set to the atmospheric value and the velocity is evaluated from the internal side of the calculation domain [53]. In WindSim, the default standard $\kappa-\varepsilon$ model is used for turbulence closure with possible modification choices such as $\kappa-\varepsilon$ RNG [19] Yap correction [54] and low Reynolds number models for a better simulation of flow separation. Because WindSim solves the conservation of energy equation, it is in principle capable of simulation motions induced by a temperature gradient. WindSim uses the so-called body-fitted coordinates (BFC) with refinement towards the ground and it is possible to perform nested runs with increasingly detailed resolution where inner boundary conditions are taken from larger run results. Results from meso-scale NWP models can also be integrated in Wind-Sim to improve boundary inlet conditions. One drawback is that WindSim is computationally more demanding than Meteodyn MT.

3.4.2   Validation of CFD models

It is speculated by many that CFD models will be the next generation of wind flow modelling for wind energy siting purposes, owing to the fast growing computational power of personal computers. Great effort has been spent recently on validating various CFD wind flow models, usually against WAsP, the industrial standard, and by conducting wind tunnel experiments. Wind tunnel experiments have shown that CFD models can simulate well the flow patterns around idealised escarpments and steep hills, even on the leeside where recirculation occurs [55]. Many case studies using wind data from field experiments such as the Askervein hill project have shown good agreement with measurement on the windward slope and at the hill top [56]. However, overall, the tests of CFD models against WAsP have not reached a consensus to favour one over the other. Some tests claim significant improvements [57] over WAsP while others have reached the opposite conclusion. CFD models may have only a small edge over short distances when the micro-scale wind mechanism dominates, but produce significantly larger errors than WAsP over longer distance when meso-scale phenomena start to impact [58].

In order to validate different wind flow models, Risø National Laboratory (Denmark) carried out an extensive measurement campaign over the small isolated island of Bolund. During the campaign, velocity and high-frequency turbulence data were collected simultaneously from 35 anemometers distributed on 10 masts, thereby generating a large database designed to validate CFD codes [59]. A blind comparison of various CFD models based on the measurements showed that the mean speed-up error of the RANS two-equation models was 13.6%, while the best of them had an error of 10.2%, even though these models with two-equation $\kappa-\varepsilon$ turbulence closures generally gave the best results among others [60]; the average overall error in predicted mean wind speed of the top 10 CFD-based models was on the order of $13-17\%$ for the principal wind directions [61]. The error on turbulence kinetic energy was much larger (about 22% error given by the best prediction). The errors are still too large to be ignored by any wind farm developer. The models used in the blind comparison were theoretically very similar to each other, but the results were scattered, a serious concern as it indicated high sensitivity to user input [62].

CFD wind flow modelling still cannot claim success over linear models and it is often questioned by many as to whether more sophisticated models actually produce more accurate wind resource estimates [46]. The validation of CFD results is still a game going on.

3.5 Meso Scale NWP Models

The last cascade of wind flow modelling is the prognostic meso-scale numerical weather prediction (NWP) models. They solve the RANS equations as well, like CFD models, but they also describe atmospheric equations (see Section 3.1.2) as accurately as possible (see Figure 3.5). A key component of the atmospheric equations is the dynamical feedback from the physical processes like evaporation, condensation, cloud microphysics, radiation, and especially turbulence, which covers such a wide range of spatial scales that it cannot be fully simulated even in the finest and most expensive CFD models.

The NWP models are capable of simulating the time evolution of the atmospheric system and do not assume an equilibrium state of the wind, as it never does in reality due to the constant flow of energy into and out of the region. Because NWP models explicitly incorporate thermally driven forces, they are able to simulate processes such as land–sea breezes, downslope winds, mountain–valley circulation, low-level jets, etc., which in turn help make sense of phenomena that are hard to explain otherwise. While these models are 'realistic' and capable of simulating wind flows in complex terrain, they are expensive to operate, require extensive computer facilities and need expertise for their operation. These models include KAMM [63], MM5 [64], RAMS [65], WRF and many more.

Meso-scale NWP models usually employ a nested grid. Outputs produced on the larger domain are used as boundary conditions on the smaller grid [66]. The typical finest resolution of mesoscale NWP models is in the magnitude of $1 \times 1 \text{ km}^2$, which is too coarse to resolve relevant micro-scale variations within complex terrain [48]. A few coupling methods have

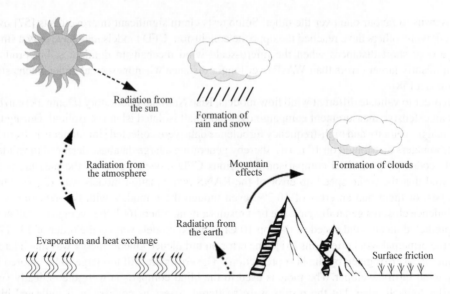

Radiation from
the sun

Formation of
rain and snow

Radiation from
the atmosphere

Mountain
effects

Formation of clouds

Radiation from
the earth

Evaporation and heat exchange

Surface friction

Figure 3.5   Types of interactions considered in an NWP model

been proposed to tackle this problem, because the results from a meso-scale model can be used to substantially improve the boundary inlet conditions of micro-scale models. One of the first was the KAMM-WAsP [67] method developed by Risø (Denmark) in the 1990s to generate a wind resource map of very high resolution. Another good example is the MASS-WindMap [68] (commercial name MesoMap) method developed and validated by AWS Truepower (United States).

The prognostic property of the NWP models allows them to generate long-term wind data time series based on over 50 years of global weather data archives. The virtual wind data series can then be statistically adjusted and corrected based on short-term precise on-site wind measurements, a process called measure–correlate–predict (MCP) (see Chapter 6). It has gradually become a common practice in wind resource assessment engineering. A few commercial organizations have started to offer meso-scale NWP wind resource maps and virtual long-term wind data time series. The wind resource map over a large region is invaluable for a site hunt (or call it macro-siting as opposed to micro-siting), but one has to be aware of the uncertainty of NWP model results (see the next section). The results are by and large a double-edged sword and have to be justified by on-site wind data as they may easily lead to completely wrong decisions.

NWP models are also an important tool for investigating extreme wind (e.g. downslope windstorms and typhoons) and weather conditions (e.g. icing and lightening probabilities) for a wind farm and mitigating the risks of such events.

Each equation in NWP models provides a tendency to provide, for example, the rate of change for one atmospheric variable, meaning that they are able to forecast wind conditions for a given wind farm in a refined time scale. The ability to predict wind energy output of a given wind farm in a few hours is important for the grid to better adapt the fluctuating wind power, especially where wind power is saturating the grid.

3.6   Inherent Uncertainties in Wind Flow Modelling

Lorenz, a famous American meteorologist, was using a numerical computer model to rerun a weather prediction when, as a shortcut on a number in the sequence, he entered the decimal 0.506 instead of entering the full 0.506127. The result was a completely different weather scenario. In later speeches and papers Lorenz used a poetic phrase 'the butterfly effect' (see Figure 3.6) to describe such phenomena [69]. Does the flap of a butterfly's wings in Brazil set off a tornado in Texas? The phrase refers to the idea that a butterfly's wings might create tiny changes in the atmosphere that may ultimately alter the path of a tornado or delay, accelerate or even prevent the occurrence of a tornado in another location; in other words, a small change at one place in a deterministic nonlinear system can result in large differences in a later state (Chaos Theory). Note that the butterfly does not power or directly create the tornado. The flapping wing represents a small change in the initial condition of the system, which causes a chain of events leading to large-scale alterations of events. One set of conditions leads to a tornado while the other set of conditions does not.

For a wind engineer, it may not be necessary to fully grasp all the complicated equations and modelling processes, but it is definitely crucial to become aware of the inherent 'butterflies' in wind flow modelling. The first 'butterfly' comes from the grid resolution. The terrain in the model's representation of the virtual world does not completely correspond to the real terrain. A different mesh method may lead to entirely different outcomes. A problem with one cell may cause the entire system to fail to converge. Secondly, errors may inevitably be induced by the discretization process, including finite difference, truncation error and accuracy of the solution of the differential equations. Furthermore, the equations may be incomplete as there are things we know but ignore, things we think we know and things we do not know. We need to keep in mind the fact that wind energy density is roughly proportional to the cubic wind speed. Errors in the wind speed will be amplified in terms of wind energy.

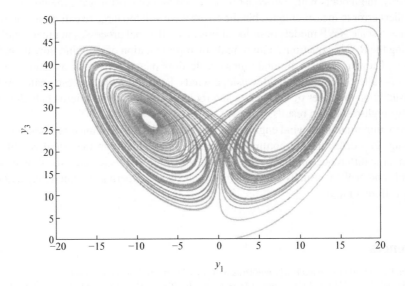

Figure 3.6   Demonstration of the butterfly effect *Source*: Wikipedia.org

More sophisticated wind flow modelling usually involves more nonlinear processes and equations, which make it more vulnerable to the chaos and present more unpredictable behaviour. Here is the essence of the argument of the simpler but faster and more predictable wind flow models against the more sophisticated but more resource demanding and user-sensitive ones. Personally, I think they should work together rather than competing against each other.

3.7 Summary

This chapter attempts to classify different numerical wind flow modelling methods and articulate their theories and pros and cons. Another useful way of classifying wind flow models is to distinguish between diagnostic and prognostic models. Diagnostic models are ones capable of reconstructing a steady-state wind field from a set of initial experimental data; the Jackson–Hunt models and mass-consistent models belong to this category. CFD wind flow models are usually in this category as well but are capable of running as prognostic models, which describe time-evolving three-dimensional wind fields by solving time-dependent equations, like the NWP models.

WAsP [22] is based on the Jackson–Hunt model [17], which solves linearised Navier–Stokes equations with assumptions like small terrain perturbation, attached flow and simple mixing-length turbulence closure. It works well under simple terrain conditions but overestimates the orographic speed-up effect when flow separation occurs.

The mass-consistent models only solve the conservation equation of mass. They usually give a better simulation of the orographic effect compared to WAsP without compromising computational convenience.

The CFD models solve incompressible, steady-state nonlinear RANS equations with more general turbulence closure techniques. According to the Bolund experiment conducted by Risø (Denmark), the models with two-equation $\kappa-\varepsilon$ turbulence closure gave the best results with the smallest error to measurements, but the errors were still too large to call it satisfactory [60].

The meso-scale NWP models describe atmospheric thermal processes in great detail as well as solving the RANS equations, a more realistic representation of the wind flow in the boundary layer. They are able to simulate meso-scale flow phenomena such as land–sea breeze, mountain–valley circulation and downslope winds. Resolution of the meso-scale NWP models is usually on the order of kilometres and requires coupling with micro-scale flow models to resolve high-resolution results.

It is very important for a wind engineer to understand the inherent uncertainties of wind flow modelling. In a deterministic nonlinear system, a small change in the initial conditions can result in large differences in the results. This is why more sophisticated wind flow modelling, such CFD and NWP, is usually very sensitive to the user's experience, which may end up with slightly different input.

References

[1] John, D.A. (1995) *Computational Fluid Dynamics*, 1st edition, McGraw-Hill Science.
[2] Falkovich, G. (2011) *Fluid Mechanics: A Short Course for Physicists*, Cambridge University Press, Cambridge.
[3] Stull, R.B. (1988) *An Introduction of Boundary Layer Meteorology*, Springer.

[4] Freudenreich, K., Kaiser, K., *et al.* (2004) Reynolds Number and Roughness Effects on Thick Airfoils for Wind Turbines. *Wind Energy*, **28** (5), 529–546.

[5] Pope, S.B. (2000) *Turbulent Flows*. Cambridge University Press, Cambridge.

[6] Reynolds, O. (1895) On the Dynamical Theory of Incompressible Viscous Fluids and the Determination of the Criterion. *Proceedings of the Royal Society of London*, **451**.

[7] Prandtl, L. (1925) Uber die ausgebildete Turbulenz. *Zeitschrift fur Angewandte Mathematik und Mechanik*, **5**, 136–139.

[8] Murakami, S. (1988) Overview of Turbulence Models Applied in CWE-1997. *Journal of Wind Engineering and Industrial Aerodynamics*, **74–76**, 1–14.

[9] Yakhot, V., Orzag, S.A. (1986) Renormalisation Group Analysis of Turbulence: Basic Theory. *Journal of Science and Computing*, **1**, 3–51.

[10] Patel, V.C., Rodi, W., Scheuerer, G. (1985) Turbulence Models for Near Wall and Low Reynolds Number Flows: A Review. *American Institute of Aeronautics and Astronautics*, **23**, 1308–1319.

[11] Apsley, D.D. (1995) Numerical Modelling of Neutral and Stably Stratified Flow and Dispersion in Complex Terrain. University of Surrey, UK.

[12] Rahman, M.M., Siikonen, T. (2002) Low Reynolds Number $\kappa-\varepsilon$ Model with Enhanced Near-Wall dissipation. *American Institute of Aeronautics and Astronautics*, **40** (7), 1462–1464.

[13] Esau, I. (2004) Simulation of Ekman Boundary Layers by Large Eddy Model with Dynamic Mixed Subfilter Closure. *Environmental Fluid Mechanics*, **4**, 273–303.

[14] Piomelli, U. (1999) Large-Eddy Simulation: Achievements and Challenges. *Progress in Aerospace Science*, **35**, 335–362.

[15] Landberg, L., Myllerup, L., Rathman, O., Petersen, E.L., Jorgensen, B.H., Badger, J., Mortensen, N.G. (2003) Wind Resource Estimation – An Overview. *Wind Energy*, **6**, 261–271.

[16] Bechmann, A. (2012) *WAsP CFD – A New Beginning in Wind Resource Assessment*, Risø National Laboratory, Denmark

[17] Jackson, P.S., Hunt, J.C.R. (1975) Turbulent Wind Flow over Low Hill. *Quarterly Journal of the Royal Meteorological Society*, **101**, 929–955.

[18] Hunt, J.C.R., Leibovich, S., Richards, K.J. (1988) Turbulent Shear Flow over Low Hills. *Quarterly Journal of the Royal Meteorological Society*, **114**, 1435–1470.

[19] Hunt, J.C.R., Richards, K.J., Brighton, P.W.M. (1988) Stratified Shear Flow over Low Hills. *Quarterly Journal of the Royal Meteorological Society*, **114**, 859–886.

[20] Carruthers, D.J., Holroyd, R.J., Hunt, J.C.R., *et al.* (1991) UK Atmospheric Dispersion Modelling System. *Proceedings of the 19th NATO/CCMS International Technical Meeting on Air Pollution Modelling and its Application*.

[21] Mason, J.R., Sykes, I. (1979) Flow over an Isolated Hill of Moderate Slope. *Quarterly Journal of the Royal Meteorological Society*, **105** (444), 383–395.

[22] Mortensen, N.G., Landberg, L., Troen, I., and Petersen E.L. (1993) *WAsP Wind Atlas Analysis and Application Program, User's Guide*. Risø National Laboratory, Denmark.

[23] Beljaars, A.C.M., Walmsley, J.L., Taylor, P.A. (1987) A Mixed Spectral Finite-Difference Model for Neutrally Stratified Boundary-Layer Flow over Roughness Changes and Topography. *Boundary-Layer Meteorology*, **38**, 273–303.

[24] Resoft. http://www.resoft.co.uk/English/html/windflow.htm (last accessed 16 December 2013).

[25] Xu, D., Taylor, P.A. (1992) A Nonlinear Extension of the Mixed Spectral Finite Difference Model for Neutrally Stratified Turbulent Flow over Topography. *Boundary-Layer Meteorology*, **59**, 177–186.

[26] Corbett, J.F. (2007) RAMSIM – A Fast Computer Model for Mean Wind Flow over Hills. Risø National Laboratory, Risø-PhD-17(EN), Denmark.

[27] Troen, I. (1990) A High Resolution Spectral Model for Flow in Complex Terrain. *Proceedings from the 9th Symposium on Turbulence and Diffusion*, Denmark.

[28] Troen, I., Petersen, E.L. (1989) *European Wind Atlas*, Risø National Laboratory, Denmark.

[29] Troen, I., de Baas, A.F. (1986) A Spectral Diagnostic Model for Wind Flow Simulations in Complex Terrain. *Proceedings of the EWEC '86 European Wind Energy Association Conference and Exhibit*, Rome, pp. 243–250.

[30] Walmsley, J.L., Troen, I., Lalas, D.P., Mason, P.J. (1990) Surface-Layer Flow in Complex Terrain: Comparison of Models and Full-Scale Observations. *Boundary-Layer Meteorology*, **52**, 259–281.

[31] WAsP 9 Help Facility and On-line Documentation, Risø National Laboratory, Denmark

[32] Mortensen, N.G., Petersen, E.L. (1997) *Influence of Topographic Input Data on the Accuracy of Wind Flow Modelling in Complex Terrain*, Risø National Laboratory, Denmark.

[33] Bowen, A.J., Mortensen, N.G. (1996) Exploring the Limits of WAsP, The Wind Atlas Analysis and Application Program. *European Union Wind Energy Conference*, Sweden.

[34] Bowen. A.J., Mortensen, N.G. (2004) *WAsP Prediction Errors Due to Site Orography*, Risø National Laboratory, Denmark.

[35] Lange, B., Hojstrup, J. (2001) Evaluation of the Wind Resource Estimation Program WAsP for Offshore Applications. *Journal of Wind Engineering and Industrial Aerodynamics*, **89**, 271–291.

[36] Mortensen, N.G., Bowen, A.J., Antoniou, I. (2006) *Improving WAsP Predictions in (Too) Complex Terrain*. Riso National Laboratory, Denmark.

[37] Restive, A. (1991) Resource Assessment in Regions of Portugal with Complex Terrain. *Proceedings of the E.C. Wind Energy Conference*, Holland, pp. 797–801.

[38] Traci, R.M., Phillips, G.T., Patnaik, P.C., Freeman, B.E. (1997) Development of a Wind Energy Site Selection Methodology. NTIS U.S. Department of Energy, Technical Report RLO/2440-11.

[39] Phillips, G.T. (1979) A Preliminary Users Guide for the NOABL Objective Analysis Code. NTIS U.S. Department of Energy, Technical Report DOE Contract AC06-77/ET/20280.

[40] Sherman, C.A. (1978) A Mass Consistent Model for Wind Field over Complex Terrain. *Journal of Applied Meteorology*, **17**, 312–319.

[41] Rodrigues, D.J., Greenly, G.D., Gresho, P.M., Lange, R., Lawver, B.S., Lawson, L.A., Walker, B. (1982) *User's Guide to the MATHEW/ADPIC Models*, Lawrence Livermore National Laboratory, University of California, Atmospheric and Geophysical Sciences Division, Livermore, California UASG 82-16.

[42] Geai, P. (1987) Méthode d'interpolation et de reconstitution tridimensionelle d'un champ de vent: le code d'analyse objective MINERVE, Report EDF/DER, HE/34.

[43] Endlich, R.M., Ludwig, F.L., Bhumralkar, C.M., Estoque, M.A. (1982) A Diagnostic Model for Estimating Winds at Potential Sites for Wind Turbines. *Journal of Applied Meteorology*, **21**, 1442–1454.

[44] Ratto, C.F., Festa, R., Nicora, O., Mosiello, R., Ricci, A., Lalas, D.P., Frumento, O.A. (1990) Wind Field Numerical Simulations: A New User-Friendly Code. *1990 European Community Wind Energy Conference*, Madrid, Spain, pp. 130–134.

[45] Brower, M.C. (1999) Validation of the WindMap Program and Development of MesoMap. *Proceedings from AWEA's WindPower Conference*, Washington, DC, USA.

[46] Philippe, B., Brower, M.C. (2012) *Wind Flow Model Performance – Do More Sophisticated Models Produce More Accurate Wind Resource Estimates?* AWS Truepwer, LLC, USA.

[47] Download openWind at http://awsopenwind.org/.

[48] Strack, M., Riedel, V. (2004) State of the Art in Application of Flow Models for Micrositing. *DEWEK 2004 - Proceedings*, available at: http://www.dewi.de/dewi/fileadmin/pdf/publications/Publikationen/s07_4_strack.pdf.

[49] Mogensen, S.H., Hristov, Y.V., Knudsen, S.J., Oxley G.S. (2012) Validation of CFD Wind Resource Mapping in Complex Terrain Based on WTG Performance Data. 2012 *Proceedings of the EWEA Conference*.

[50] Technical Note-Meteodyn WT.http://meteodyn.com/wp- content/uploads/2012/06/Technical_note_WT2.2.pdf (last accessed 20 December 2013).

[51] Yamada, T. (1983) Simulations of Nocturnal Drainage Flows by a $q^2 l$ Turbulence Closure Model. *Journal of the Atmospheric Sciences*, **40**, 91–106.

[52] Arritt, R.W. (1987) The Effect of Water Surface Boundary Layers. *Boundary-Layer Meteorology,* **40**, 101–125

[53] Daniele, F. (2007) *Wind Energy Resource Evaluation in a Site of Central Italy by CFD Simulations*. D.I.Me.Ca, Dipartimento di Ingegneria Meccanica, Cagliari, Italy.

[54] Yap, C. (1987) *Turbulent Heat and Momentum Transfer in Recirculating and Impinging Flows*. PhD Thesis, Faculty of Technology, University of Manchester, UK.

[55] Bitsuamlak, G.T, Stathopoulos, T., Bédard, C. (2004) Numerical Evaluation of Wind Flow over Complex Terrain: Review. *Journal of Aerospace Engineering*, **17**, 135–145.

[56] Kim, H., Patel, V., Lee, C. (2000) Numerical Simulation of Wind Flow over Hilly Terrain. *Journal of Wind Engineering Industrial Aerodynamics*, **87**, 45–60.

[57] Pereira, R., Guedes, R., Santos, C.S. (2010) Comparing WAsP and CFD Wind Resource Estimates for the 'Regular' User. *Proceedings of the EWEC Conference*, Poland.

[58] Beaucage, P., Brower, M.C. (2011) Evaluation of Four Numerical Wind Flow Models for Wind Resource Mapping. Submitted to the related special issue of *Wind Energy Journal*. Available at: http://www.ewea.org/events/workshops/wp-content/uploads/2013/06/EWEA-RA2013-Dublin-2-1-Michael-Brower-AWS-Truepower.pdf.

[59] Bechmann, A., Berg, J., Courtney, M., Jørgensen, H., Mann, J., Sørensen, N. (2009) *The Bolund Experiment: Overview and Background*. Risø DTU, Technical Report R-1658(EN), Denmark.

[60] Bechmann, A., Sørensen, N.N., Berg, J., Mann, J. (2011) The Bolund Experiment, Part II: Blind Comparison of Microscale Flow Models. *Boundary-Layer Meteorology*, **141**, 245–271.

[61] Summer, M., Sibuet, W.C., Masson, C. (2010) CFD in Wind Energy: The Virtual, Multiscale Wind Tunnel. *Energies*, **3**, 989–1013.

[62] Manning, J., Woodcock, J., Corbett J.-F., *et al.* (2010) *Validation and Challenges of CFD in Complex Terrain for Real World Wind Farms*. GL Garrad Hassan, Bristol, UK.

[63] Adrian, G. (1987) Determination of the Basic State of a Numerical Mesoscale Model from Operational Numerical Weather Forecast. *Beitr. Phys. Atmosph.*, **60**, 3651–3670.

[64] Dudhia, J. (1993) A Nonhydrostatic Version of the Penn State-NCAR Mesoscale Model: Validation Tests and Simulation of an Atlantic Cyclone and Cold Front. *Quarterly Journal of the Royal Meteorological Society*, **121**, 1493–1511.

[65] Pielke, R.A., Cotton, R.W., Walko, R.L., Tremback, C. J., Lyons, W.A., Grasso, D.L., Nicholls, M.E., Moran, M.D., Wesley, D.A., Lee, T.J., Copeland, J.H. (1992) A Comprehensive Meteorological Modelling System – RAMS. *Meteorological Atmospheric Physics*, **49**, 69–91.

[66] Sandro, F., *et al.* (1997) *Wind Flow Models over Complex Terrain for Dispersion Calculation. Cost Action 710 – Pre-processing of Meteorological Data for Dispersion Models*, Report of Working Group.

[67] Frank, H. P., Rathmann, O., Mortensen, N.G., Landberg, L. (2001) *The Numerical Wind Atlas – The KAMM/WAsP Method*. Risø National Laboratory, Denmark.

[68] Brower, M.C, Bailey, B., Zack, J. (2001) Applications and Validations of the MesoMap Wind Mapping System in Different Wind Climates. *Windpower 2001 Proceedings*. AWEA.

[69] Lorenz, E.N. (1963) Deterministic Nonperiodic Flow. *Journal of the Atmospheric Sciences*, **20** (2), 130–141.

4

Wind Park Physics and Micro-siting

A wind park or a wind power plant usually consists of many wind turbines that are strategically arranged into a layout to maximise energy production with as minimum as possible loads on the wind turbines. The flow interaction between the wind and a wind turbine and among neighbouring wind turbines, that is wind turbine wake effects, governs the distance between the wind turbines and their layout at the end. The topic will be introduced in Section 4.3. Before looking at the wind park, it is helpful to understand the physics of wind energy conversion; after all, the goal of a wind resource assessment is to calculate the amount of wind energy convertible by wind turbines. Section 4.2 will also help us gain a basic understanding of wind turbine aerodynamics.

4.1 Wind Power Density

Wind energy is the kinetic energy of air mass. It can be expressed as

$$K = \frac{1}{2}mv^2 \tag{4.1}$$

where K is the kinetic energy for an air mass m with a wind speed of v. The air mass m can be derived from the production of its density ρ and volume V. The air mass passing through a normal section area of A during a time period of t at constant wind speed of v is

$$m = \rho V = \rho A v t \tag{4.2}$$

Therefore, the kinetic energy of the air mass is

$$K = \frac{1}{2}\rho v^3 A t \tag{4.3}$$

Wind power density characterises the ability of the kinetic energy being converted into power; the higher the wind power density, the more economic potential the wind park has. It can be used as the initial guidance in selecting regions for wind power projects. Wind power

Wind Resource Assessment and Micro-siting, Science and Engineering, First Edition. Matthew Huaiquan Zhang.
© 2015 John Wiley & Sons, Ltd. Published 2015 by John Wiley & Sons, Ltd.

density P can be defined as the kinetic energy of the air mass per unit area per second ($A = 1$; $t = 1$); thus

$$P = \frac{1}{2}\rho v^3 \tag{4.4}$$

Equation (4.4) is when the wind speed v is constant, but in reality the wind speed continuously changes. Statistically, the mean wind speed follows a Weibull distribution (see Chapter 5 for details); thus

$$v_{ave} = a\Gamma\left(1 + \frac{1}{k}\right) \tag{4.5}$$

where k and a are the shape and scale parameters of the Weibull distribution and v_{ave} the annual mean wind speed.

Replacing v in Equation (4.4) by v_{ave} in (4.5), we get

$$P = \frac{1}{2}\rho a^3 \Gamma\left(1 + \frac{3}{k}\right) \tag{4.6}$$

If the Weibull distribution parameters are unknown, as an approximation the Rayleigh distribution, that is when $k = 2$, can be assumed to calculate the wind power density:

$$P = \frac{3}{8}\sqrt{\pi}\rho a^3 = \frac{3}{\pi}\rho v_{ave}^3 \tag{4.7}$$

Example 4.1

A prospective wind park has an estimated annual mean wind speed of 7.8 m/s and air density of 1.15 kg/m³. What is the wind power density of the wind park?

Solution

Because the Weibull distribution parameters are unavailable, Equation (4.7) should be used to estimate the wind power density. Therefore,

$$P = \frac{3}{\pi}\rho v_{ave}^3 = \frac{3}{\pi} \times 1.15 \times 7.8^3 = 521.4 \text{ W/m}^2 \tag{4.8}$$

Example 4.2

For the same wind park as above, we have now found that the k parameter of the Weibull distribution is 2.3. What is the wind power density now?

Solution

According to (4.5):

$$a = \frac{v_{ave}}{\Gamma\left(1 + \frac{1}{k}\right)} = \frac{7.8}{\Gamma\left(1 + \frac{1}{2.3}\right)} = 8.8 \text{ m/s} \tag{4.9}$$

This time, Equation (4.6) is chosen for a more accurate calculation; thus

$$P = \frac{1}{2} \times 1.15 \times 8.8^3 \times \Gamma\left(1 + \frac{3}{2.3}\right) = 458 \text{ W/m}^2 \qquad (4.10)$$

It should be noted that there is a 13.8% discrepancy between Equations (4.8) and (4.10), which means that the mean wind speed alone may not be enough to accurately evaluate the economic potential of a wind park, as wind power density directly determines cost efficiency in using wind energy.

Wind power density may also be calculated directly from wind data records following

$$P = \frac{1}{2n} \sum_{i=1}^{n} \rho_i v_i^3 \qquad (4.11)$$

where n is the number of records in the period, v_i is the ith recorded wind speed in m/s and ρ_i is the corresponding air density in kg/m^3. The calculation of air density is described in Chapter 9.

Wind power density indicates the wind energy potential of the site. It may be categorised as poor (<250 W/m^2), fair (250~400 W/m^2), good (400~600 W/m^2) or excellent (>600 W/m^2) at hub height. Additional information, such as topography, grid connection, road connection and electricity price, should also be considered to determine the economic wellbeing of the prospective wind park project.

4.2   Wind Power Conversion

4.2.1   Betz's Limit

Betz's law calculates the maximum possible energy that can be extracted from the wind in open flow regardless of the design of a wind turbine. It was published by the German physicist Albert Betz in 1919 (see Betz, 1966 [1]). According to Betz's law, no turbine can capture more than 16/27 (59.3%) of the kinetic energy in wind.

The Betz limit is a theoretical value. Its derivation requires a few assumptions:

1. The rotor is an ideal plate without a hub and with zero mass.
2. An infinite number of blades with zero drag and any resulting drag would only lower this idealised value.
3. The flow into and out of the rotor is axial.
4. Incompressible flow with constant air density remains and there is no heat transfer between the rotor and the flow.

In Figure 4.1, A_1 and v_1 are the section area and wind speed of the upstream flow before impact, A and v the area of the rotor and the wind speed on it and A_2 and v_2 the stabilised section area and downstream wind speed after impact.

Applying the conservation of mass, the mass flow rate (the mass of air flowing per second) is given by

$$m = \rho A_1 v_1 = \rho A v = \rho A_2 v_2 \qquad (4.12)$$

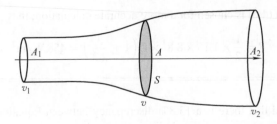

Figure 4.1 Betz's tube indicating the wind flow changes before and after hitting the rotor (S)

The force exerted on the wind by the rotor can be written as

$$F = m\frac{dv}{dt} = \rho Av(v_1 - v_2) \tag{4.13}$$

Thus, the work done per second by the force is

$$P = Fv = \rho Av^2(v_1 - v_2) \tag{4.14}$$

However, the work done per second by the force can be calculated alternatively according to the conservation of energy:

$$P = \frac{1}{2}m(v_1^2 - v_2^2) = \frac{1}{2}\rho Av(v_1^2 - v_2^2) \tag{4.15}$$

Thus

$$\rho Av^2(v_1 - v_2) = \frac{1}{2}\rho Av(v_1^2 - v_2^2) \tag{4.16}$$

Simplifying the (4.16) and we get:

$$v = \frac{1}{2}(v_1 + v_2) \tag{4.17}$$

Returning (4.17) to (4.15) yields the power extracted from the wind:

$$P = \frac{1}{4}\rho Av_1^3 \left[1 - \left(\frac{v_2}{v_1}\right)^2 + \left(\frac{v_2}{v_1}\right) - \left(\frac{v_2}{v_1}\right)^3\right] \tag{4.18}$$

Power extracted reaches the maximum value when $v_2/v_1 = 1/3$. Substituting it in (4.18) gives

$$P = \frac{16}{27} \times \frac{1}{2}\rho Av_1^3 \approx 0.593P_0 \tag{4.19}$$

where P_0 represents the total power available in the wind flow.

The Betz result seems a little hard to understand at first glance, because a speed ratio of 1/3 implies that air passing through the rotor has lost 88.9% of its kinetic energy. This seeming inconsistency can be explained by the 'milk bottle' shaped flow shown in Figure 4.1. The upstream flow has a cross-sectional area less than the rotor area, indicating that the flow begins to lose energy as it approaches the rotor and expands to exactly the rotor area as it passes through.

The existence of the maximum value makes sense intuitively as well. If the wind turbine extracted 100% of the total kinetic energy of the wind, the wind speed immediately behind the rotor would be zero or in a standstill, which would consequentially block any more wind flow from entering the rotor. Therefore, part of the flow would have to expand radially and go around the rotor, taking some kinetic energy with it. Modern utility-scale wind turbines normally peak at 75% to 80% of the Betz limit at lower wind speeds, whilst the overall efficiency falls by 15~20%.

4.2.2   Power Coefficient

The energy efficiency of a wind turbine is lower than the Betz limit. It is determined by factors including the profile and setup of the blades, pitch strategy and rotor speed. We can use the C_P coefficient to denote the percentage power extracted from the total kinetic energy of the wind and easily yield

$$P = \frac{1}{2}\rho A v^3 C_P \tag{4.20}$$

and thus achieve

$$C_P = \frac{2P}{\rho A v^3} \tag{4.21}$$

The wind speed starts to decrease before reaching the rotor. The parameter a can be defined as the fractional wind speed reduction along the rotor surface normal and thus

$$a = \frac{v_1 - v}{v} \tag{4.22}$$

where v_1 and v represent the free wind speed upstream and the wind speed at the rotor, as shown in Figure 4.1.

According to the conservation of momentum, the wind speed loss must have been transferred to the rotor. Applying a similar fluid dynamic procedure of deriving the Betz limit mentioned above, the C_P coefficient can alternatively be written as

$$C_P = 4a(1 - a)^2 \tag{4.23}$$

C_P achieves the maximum value of 16/27 when $a = 1/3$, another way of deriving the Betz limit. The result indicates that an ideal rotor can reduce the free wind speed by 2/3. thereby reaching its maximum energy efficiency.

4.2.3   Thrust Coefficient

C_P helps us to understand the rotor efficiency from the perspective of energy conservation, but it is actually more useful for us to look at it as the momentum loss of the wind due to the rotor, as the conservation of momentum formulates the most critical equations for wind flow modelling (see Chapter 3).

We can define the thrust coefficient, C_T, as the fraction of momentum absorbed by the wind turbine rotor from the free wind. The aerodynamic thrust, T, exerted by the wind is

$$T = \frac{1}{2}\rho A v^2 R C_T \tag{4.24}$$

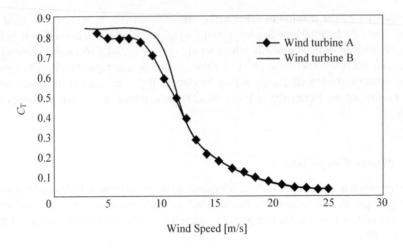

Figure 4.2   Thrust coefficient C_T curve of typical pitch-regulated wind turbines

where R is the rotor radius. Thus, the thrust coefficient C_T of the wind turbine rotor at a given wind speed is

$$C_T = \frac{2T}{\rho A v^2 R} \tag{4.25}$$

Similarly, we can derive the alternative C_T coefficient using fluid dynamic principles:

$$C_T = 4a(1-a) \tag{4.26}$$

The C_T curve of a wind turbine is one of the most important specifications of the wind turbine in terms of wind resource assessment as it has a noticeable impact on wind turbine wakes. Figure 4.2 shows the C_T curves of typical pitch-regulated wind turbines.

4.2.4   Wind Turbine Power Curve

The power curve summarises the designed electrical power output of a wind turbine at different wind speeds, established by the wind turbine manufacturers following published guidelines, most famously the IEC 61400-12-1: *Power Performance Measurements of Electricity Producing Wind Turbines* [2]. It is the key ingredient for calculating the wind turbine's energy production. We need to introduce the electrical power coefficient, C_e, to represent the amount of wind kinetic energy converted into electrical power by the wind turbine. It is not possible to convert all the kinetic energy absorbed by the rotor into actual electricity output in reality due to limited system efficiency and thus we have

$$C_e = C_p \eta_m \eta_e \tag{4.27}$$

where η_m represents the mechanical efficiency of converting the blade kinetic energy into electrical energy, determined mainly by the friction in the drive chain, especially the gearbox and the aerodynamic efficiency of the blades and η_e is the electrical efficiency of the generator. At rated power, η_m usually ranges from 0.95 to 0.97 while η_e is around 0.97~0.98.

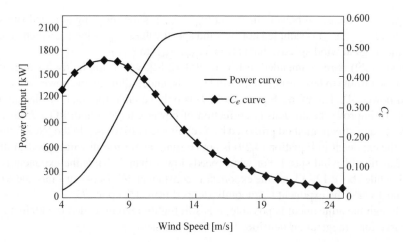

Figure 4.3   Power curve and C_e curve of a typical pitch-regulated wind turbine at air density of 1.225 kg/m³

The wind turbine power curve can be written as

$$P_e = \frac{1}{2}\rho A v^3 C_e \qquad (4.28)$$

The power curve and C_e curve of a typical pitch-regulated wind turbine is given in Figure 4.3. The pitch angle of the blades is usually maintained at its optimum operation in the lower wind speed region to maximize energy capture but is controlled to level off the power output when the wind speed exceeds the rated power wind speed [3].

The peak/rated power for a wind turbine generator is very different from that for a fuel-powered generator, in that the wind speed at which a wind turbine generates its peak power occurs for only a very small percentage of the time in most cases. Thus, one cannot make a simple comparison between one wind turbine and another using the peak power of power curves. Another important feature of the power curve is the cut-in and cut-out wind speeds. The wind turbine starts generating power when the wind speed rises above the cut-in wind speed and stops when the wind speed exceeds the cut-out wind speed to protect the wind turbine from overloading. Depending on the design, the cut-in wind speed of modern wind turbines is usually at around 3~4 m/s.

4.2.5   Power Curve Adjustment

Normally, the power curve may only be available for the desired air density level, and thus it has to be adjusted to fit the site specific air density. The adjustment can be made using the cubic relation between wind speed and power output (see Equation (4.4)). Instead of scaling the power values directly, which would alter the rated power, the adjustment is made by scaling the wind speeds according to

$$v_{site} = v_0 \left(\frac{\rho_0}{\rho_{site}} \right)^{1/3} \qquad (4.29)$$

where (v_0, ρ_0) is the given power curve and (v_{site}, ρ_{site}) is the resulting adjusted site specific power curve; (v_{site}, ρ_{site}) is sampled at new wind speed values v_{site}, which can then be interpolated at the original wind speeds, that is from (v_{site}, ρ_{site}) to (v_0, ρ_{site}).

Equation (4.29) is recommended in IEC 61400-12 [2], which implicitly assumes the efficiency of the turbine to be constant at all wind speeds, underadjusting the power values around the rated power [4]. Therefore, Equation (4.29) is only suitable for small air density corrections [4]. To improve the air density correction of power curves, WindPRO 2.7 has adopted an improved adjustment method proposed by Svenningsen (2010) [4]. In the 'New WindPRO' method, the exponent in Equation (4.29) is not constant at 1/3 for all wind speeds; rather it is made a function of wind speed. For wind speeds lower than 7–8 m/s the exponent is a constant 1/3, while above 12–13 m/s the exponent is constant at 2/3. Between 7 m/s and 8 m/s and 12 m/s and 13 m/s, the exponent is smoothly stepped from 1/3 up to 2/3.

Even though the adjustment is possible, it is still highly recommended to obtain a certified power curve for a range of air densities from the manufacturer.

4.3 Wind Turbine Wake Effects

Wind turbines extract kinetic energy from the wind to generate electricity. As a result, the kinetic energy in the wind after passing through the wind turbine must be reduced (i.e. reduced wind speed) and changed in pattern (i.e. increased turbulence intensity) according to the conservation of energy and momentum [5], like a shadow of the wind turbine reflected on the wind, called the wind turbine wake effect. This is one of the most significant sources of production loss in wind power plants. Wind turbine layout optimization is essentially a procedure of minimizing the wake effect.

Just about every discussion about the wind turbine wake effect now has to mention the famous image of the Horn's Rev offshore wind farm in Europe's North Sea (see Figure 4.4). The disturbance caused by the wind turbines upstream shoves the saturated airflow off equilibrium, resulting in water vapour condensation. As we can see from the picture, the wake effect one turbine has on the next, and the next, and on down the line can be quite dramatic. The wind turbines furthest downstream from the wind may produce over 20% less energy, or even no energy at all as they struggle to operate in intense turbulences, prematurely wearing down. The Horn's Rev offshore wind farm is particularly interesting for research because terrain effects are not important here and the wind turbines are arranged into an orderly matrix with 8 rows (east–west) and 10 columns (north–south) [6], providing extensive data from the SCADA system and met masts.

Partly due to the complexity of the wake, the two concerns of wake, that is wind speed loss and turbulent intensity augmentation, are usually modelled separately. Wind farms being developed today are larger and often in complex terrain, close to forests or offshore. Thus there is a need to re-examine the performance of the simple analytical wake models developed in the 1980s.

This chapter will only introduce wake effects of a horizontal axis wind turbine.

4.3.1 Analytical Structure of Wake

As wind turbines are becoming increasingly larger in rotor area, understanding the aerodynamic nature and the characteristics of wake effects becomes very important. Only then is it

Figure 4.4 Wake of the Horn's Rev offshore wind park *Source*: Vattenfall, photograghed by Christian Steiness

possible to find the optimal wind turbine layout in order to maximise economic yield of the wind power investment [7].

Werl (2008) [8] presents an improved analytical wake model for a horizontal axis wind turbine with an illustrative structure (see Figure 4.5) that divides the wake into three zones: the near wake, the intermediate wake and the far wake. This wake structure can be explained by applying the conservation of mass and momentum [9].

The near wake zone is about 2 rotor diameters. The airflow expands to the blade's diameter as it approaches the rotor (applying the Betz limit) causing pressure, P, to build up in front of the

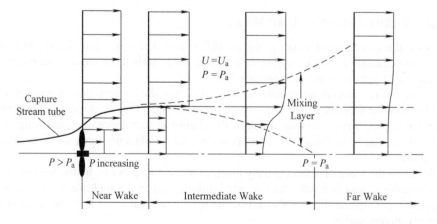

Figure 4.5 Illustration of the wake structure of a horizontal-axis wind turbine *Source*: Flodesign Inc.

rotor. The pressure plummets suddenly at the other side of the rotor as much of the air mass has been blocked off by the rotor, but increases constantly in the near wake zone due to turbulent mixing until it recovers to the free wind value, P_a. The flow velocity decreases behind the rotor as well as a result of the extraction of flow momentum by the rotor, but decreases even further in the near wake as the radius of the near wake further expands, as implied by the conservation of mass principle. The intermediate wake zone is defined by the mixing layer constantly merging the vortex wake flow and the free wind flow. The length of the intermediate wake is about 2–3 rotor diameters and ends when the mixing layer reaches the axial line and starts changing the centreline velocity. The length of the far wake is about 5 rotor diameters. It is characterised by the constant recovery of flow velocity at the centreline until it reaches the free wind value.

Applying the properties of each zone, it is possible to choose the optimal distance between the wind turbines, particularly in the prevailing wind direction, to which the level of wake impact is very sensitive. In the prevailing wind direction, the next wind turbine downstream should be no nearer than the very end of the intermediate wake of the wind turbine in front. The pressure difference in the near wake results in intense turbulent mixing and it should be avoided at all costs in the prevailing wind direction. It is thus recommended to keep the wind turbine distance no less than 5 rotor diameters in the prevailing wind direction to improve production and 3 rotor diameters in the perpendicular direction as a minimum safety requirement for the wind turbines. This is for onshore wind parks under neutrally stable air conditions.

Atmospheric stability significantly affects the turbulent mixing in the boundary layer, which in turn affects the rate of momentum recovery of the wake from the free stream. Therefore, under stable conditions such as a cold climate, the wake propagates longer compared to neutral stability, and wind turbines should be spaced further apart. Another factor with a strong influence on wake lengths is the ambient turbulence intensity, which is determined by surface roughness in the atmospheric surface layer for a given atmospheric stability (see Chapter 10). This is because turbulence intensity also directly influences the mixing rate of the wake with the free stream in a similar manner as thermal stability. Therefore, wake length is essentially a function of roughness length with longer, more sustaining wake for smaller roughness lengths. The sea surface is much smoother than the land, and therefore minimum 9 rotor diameters in the prevailing wind direction and 7 rotor diameters as the safety threshold may be recommended for offshore wind farms.

4.3.2   Reduced Velocity Wake Models

Various wake models exist today. They are different in terms of approximate solution of momentum equations, turbine representation, near wake, turbulence closure, description of the boundary layer and wake superposition [10].

Analytical wake models are usually derived only from the conservation of momentum in order to calculate velocity deficit of the wake behind a rotor. They do not consider the initial expansion region, that is the near wake, or the change in turbulence intensity in the wake. Thus they have to be coupled with a turbulence model if turbulence intensity in the wake is desired [11].

Calculating the wake of a farm usually involves a two-step process: firstly, the calculation of a single wake, that is a wake for one wind turbine, and then a multiple wake, that is the superposition of several single wakes inside the wind farm taking into account the combined effect of different wakes.

4.3.2.1   Jensen Model

The Jensen model [12], also known as the PARK model, is one of the oldest wake models and is applied in commercial wind resource assessment toolkits WAsP [13], WindPRO [14] and WindFarmer [15]. It is easy to code, requiring only a short computing time, and agrees well with measurements. The Jensen model is based on the assumption that the wake diameter expands linearly with respect to the distance behind the rotor disc, as seen in Figure 4.6. The near wake zone behind the rotor is ignored in the model. The slope of the wake is defined by the entrainment constant k, the wake decay coefficient.

According to Figure 4.6, the diameter of the wake, D_W, at a given distance X behind the rotor with a diameter of D can be easily derived as

$$D_W = D + 2kX \tag{4.30}$$

Applying the conservation of momentum and manipulating Equation (4.26) gives [12] the velocity in the wake, v, at any distance of X:

$$U = U_0 \left[1 - \left(1 - \sqrt{1 - C_T} \right) \left(\frac{D}{D + 2kX} \right)^2 \right] \tag{4.31}$$

where v_0 represents the free-stream velocity and C_T is the thrust coefficient of the wind turbine.

It is clear from (4.31) that the wake at a given free wind speed is determined by the rotor diameter, D, the thrust coefficient, C_T (the aerodynamic properties and control strategy of the rotor blades), and the empirically defined decay coefficient, k, which is an indicative parameter of atmospheric stability. The value of k, the only user-defined parameter in the model, is larger for more unstable air stratification where the wake would decay quicker. Parameter $k = 0.04$ is usually assumed for offshore and $k = 0.075$ for neutrally stratified onshore conditions.

As for the superposition of multiple wakes, the incident wind speed at a downwind turbine is the free-stream wind speed minus Equation (4.31) multiplied by the fractional overlap between the downwind rotor area and the cross-section of the upwind turbine wake.

WindPRO (2.5 and above) allows the Jensen model (Jensen 2005) to be able to work together with turbulence models and suggests that a higher turbulence intensity leads to a

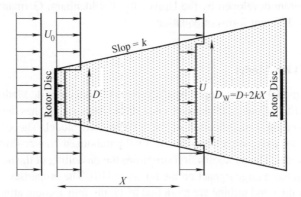

Figure 4.6   The structure of the Jensen wake model. U_0 represents the free-stream velocity, U the velocity in the wake at the distance of X behind the rotor disc with a diameter of D and D_W is the diameter of wake at distance X [12]. *Source*: Risø National Laboratory

higher wake decay coefficient as high turbulence causes the wind in the wake to mix faster with the surrounding air [16].

4.3.2.2   Larsen Model

The Larsen wake model [17], also known as the EWTS II model, is based on the Prandtl turbulent boundary layer equations for defining the wake behind the wind turbine. Prandtl's mixing length theory is used to obtain the closed-form solutions. The flow is assumed to be incompressible and stationary and axisymmetric. It is optional in the model to use a first- or second-order approximation approach. The choice of order is only relevant for the zone directly behind the rotor and thus usually has no influence on the results [18].The Larsen model is significantly sensitive to changes in turbulence intensity [19]. It is also included in WindPRO [14].

4.3.2.3   Ainslie Model

The Ainslie model (1988) [20] is a CFD model that assumes the wake to be axisymmetric. leading to fewer equations to solve in two-dimensional cylindrical coordinates. The flow is considered incompressible and inviscid without external forces or pressure gradients. Two-dimensional RANS equations together with the continuity equation are then solved numerically with eddy viscosity turbulence closure to calculate wake behaviour. This assumption of no pressure gradient is not valid in the near wake region. Hence the model has to be initialised at the end of the near wake (assumed to be two dimensional) with an empirical Gaussian wake profile as the boundary condition.

The model was extended by Lange *et al.* (2003) [21] for offshore applications by including the effects of turbulence and atmospheric stability on the wake development. The low turbulence intensity offshore leads to slower wake recovery and therefore a longer near wake region.

The Ainslie model is implemented in WindPRO [14], WindFarmer [15] and FLaP [21] (the Farm Layout Program developed by the University of Oldenburg, Germany) and openWind [22], with slightly different forms of equations.

4.3.2.4   UPMWAKE Model

The UPMWAKE model was developed by Universidad Polytecnica de Madrid, Spain (Crespo *et al.*, 1988 [23]). ECN (Energy Research Centre of the Netherlands) has developed the model WAKEFARM based on a modification of the UPMWAKE model. The UPMWAKE model is also a CFD model based on a three-dimensional parabolised Navier–Stokes code using a $\kappa-\varepsilon$ turbulence model. The parabolisation simplifies the modelling of the near wake considerably, which in turn has a large impact on the far wake [10]. The properties of the nonuniform incident flow over the wind turbine are modelled by taking into account atmospheric stability, given by the Monin–Obukhov length and surface roughness [24]. No assumptions are required regarding the type of superposition or the type of wake to be used, as all the wakes and their interactions are effectively calculated by the code [5].

4.3.2.5 Elliptic Models

Several elliptic models have been proposed to study the flow around wind turbines and through wind farms. The Robert Gordon University (RGU) has developed a fully elliptic three-dimensional Navier–Stokes solver with κ–ε turbulence closure. These models are able to capture a lot of details in the near wake, but require very intensive computational power [25]. For general micro-siting work, these models are usually not feasible and may not yield significantly better results.

4.3.3 Added Turbulence Wake Models

Another severe impact of the wake on downstream wind turbines is the increase in turbulence intensity, which is used directly for load calculations. As mentioned earlier, the analytical wake models have to be combined with turbulence models to calculate the added turbulence in the wake.

4.3.3.1 Frandsen Wake Turbulence Model

The Frandsen model [26, 27], also referred to as the DIBt (Deutsche Institut für Bautechnik) model in WindPRO for the sake of users in Germany, is probably the most widespread wake turbulence model in the industry. According to the Frandsen model, if the smallest spacing between two wind turbines in a wind farm is more than 10 rotor diameters, the wake effects in terms of added turbulence does not have to be considered. Otherwise, the added turbulence has to be calculated.

The turbulence intensity in the wake is defined as effective turbulence, I_{eff}, which is given by

$$I_{eff} = \left[\left(1 - N P_w \right) I_o^m + P_w N \sum_{i=1}^{N} I_T^m s_i \right]^{\frac{1}{m}} \tag{4.32}$$

where N is the number of neighbouring wind turbines, m is the Wöhler exponent for the applied material of the structural components, I_0 is the ambient turbulence intensity, s_i is the relative distance to the neighbouring turbine i in rotor diameters (D); P_w is the wake probability and is taken to be 0.06 corresponding to a horizontal angle of $0.06 \times 360 = 21.6$ deg (valid only if a uniform wind direction probability distribution is assumed) and I_T is the maximum turbulence intensity in the centre of the wake and is given by the root mean square of wake added turbulence intensity, I_w, and the ambient turbulence, I_0, as it is assumed that the wind turbine creates an additional turbulence kinetic energy that should be added to the ambient one:

$$I_T = \sqrt{I_w^2 + I_0^2} \tag{4.33}$$

$$I_w = \frac{1}{1.5 + 0.8 s_i / \sqrt{C_T}} \tag{4.34}$$

As seen in Figure 4.7, the number of neighbouring wind turbines, N, can be taken to be $N = 1$ for two wind turbines, $N = 2$ for one row of wind turbines, $N = 5$ for two rows and $N = 8$

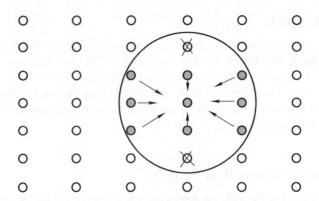

Figure 4.7   Turbines taken into account in the Frandsen model. Solid circles with an arrow pointing to the wind turbine in consideration are taken into account, while the crossed circles are not [26] *Source*: Risø National Laboratory

when there are more than two rows. Note that the wake of a turbine that is in front of another turbine is not taken into account (see the crossed circles in Figure 4.7).

For large wind farms with more than five turbines between the turbine under consideration and the edge of the wind farm, or when the turbines are spaced less than three rotor diameters from each other in the direction perpendicular to the predominant wind direction, a correction has to be made to the ambient turbulence intensity, I_0, and wake added turbulence, I_w, which are given by

$$I'_0 = \frac{1}{2}(\sqrt{I_w^2 + I_0^2} + I_0) \tag{4.35}$$

$$I'_w = \frac{0.36}{1 + 0.2\sqrt{s_d s_c / C_T}} \tag{4.36}$$

where s_d and s_c are the mean downwind and crosswind relative spacings in rotor diameters (D) in the case of a regular grid layout.

4.3.3.2   Larsen Wake Turbulence Model

The Larsen wake turbulence model is based on the assumptions that only surface and wake shear mechanisms contribute significantly to turbulence production and that these mechanisms are statistically independent and thus additive [11]. The wake added turbulence for relative spacing larger than two dimensions is given by

$$I_w = 0.29s^{-1/3}\sqrt{1 - \sqrt{1 - C_T}} \tag{4.37}$$

The total turbulence intensity in the wake can be calculated as the root mean square of I_w and I_0, which is similar to Equation (4.33).

There are other wake turbulence models [11] that are not included in this chapter, such as the Danish recommendation [28], the Lange wake turbulence model for the Ainslie velocity deficit model mentioned above, Quarton/TNO models [29] the Crespo model (Crespo and Hernández, 1996 [30], etc.

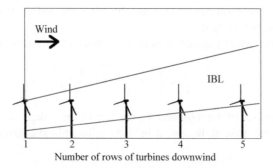

Figure 4.8   Illustration of the development of the internal boundary layer (IBL) in a deep-array wind farm, represented by the two curves

4.3.4   Deep Array Wake Models

The large scale wind farms with hundreds up to several thousands of wind turbines stretching over long distances are under development nowadays, especially for offshore wind farms. The standard wake models ignore two-way interactions between the atmosphere and the turbines and assume the ambient wind to be unaffected. They may significantly underestimate wake losses, thereby overestimating energy production in large offshore wind farms with more than five rows inside [31].

The point is that wind turbines themselves become part of the topographic features of the surface in large wind farms, which can alter the wind profile in the atmospheric boundary layer outside the zone of direct wake impact, both within and around the array, thereby reducing the amount of energy available to the turbines for power production [32]. In other words, wind turbines are part of the dynamic equilibrium of momentum in the boundary layer that are similar to trees or other roughness elements in terms of the development of the internal boundary layer (IBL) (see Section 2.1.3) [33]. As shown in Figure 4.8, two IBLs are formed after the first row of turbines: one above the rotor created by the increased roughness; the other one underneath it representing a transition back to ambient surface conditions.

The development of IBLs created by wind turbines is more pronounced offshore, where the drag prompted by the turbines and the relatively low roughness of the ocean surface makes the so-called deep-array effect especially evident. It is further enhanced by the relatively low ambient turbulence intensity offshore, which hinders the momentum recovery of the wake from the free stream above. The effect is weakened onshore but it may nonetheless be significant in projects involving hundreds to thousands of wind turbines [32].

4.3.4.1   Frandsen Model

The Frandsen model (Frandsen and co-workers, 2006, [31, 34]) links the small scale and large scale features of the flow in wind farms. The model distinguishes three different wake regimes from the upwind end of the wind farm. In the first regime, single- or multiple-wake flow is derived through an analytical link between the expansion of the wake and the asymptotic velocity deficit without interaction between wakes from neighbouring rows. The second regime starts when two neighbouring wake flows merge and the wakes are limited to only expand vertically, similar to the flow after a terrain roughness change. The wake in the third and last

regime is when the wind farm can be seen as infinitely large and the wake flow is in balance
with the atmospheric boundary layer.

In a single wake the velocity deficit is assumed to be constant and is given by

$$U = \frac{1}{2}U_0\left(1 \pm \sqrt{1 - 2\frac{A}{A_W}C_T}\right)$$

(4.38)

where A is the swept area of the rotor and A_W the area of the wake.

The expansion of the wake at distance X behind a wind turbine is given by

$$D_W = D\left(\beta^{k/2} + \alpha\frac{X}{D}\right)^{1/k}$$

(4.39)

$$\beta = \frac{1 + \sqrt{1 - C_T}}{2\sqrt{1 - C_T}}$$

(4.40)

where α is an empirical constant related to the thrust coefficient, C_T, and is set to 0.7 [10], k
is the expansion exponent and considered to be 2 or 3 and β is the initial wake expansion and
can be used to define the effective rotor diameter, D_{eff}:

$$D_{eff} = D\sqrt{\beta}$$

(4.41)

Rathmann et al. (2006) [31] gives more details regarding the method for wake combination.

4.3.4.2   open Wind DAWM Model

Brower and Robinson (2012) [32] (AWS Truepower) used the wind farm equivalent roughness
z_{00} based on the Frandsen theory [34]:

$$z_{00} = h_H \exp\left(-\frac{\kappa}{\sqrt{\frac{\pi}{8s_d s_c}C_T + \left(\frac{\kappa}{\ln\left(h_H/z_0\right)}\right)^2}}\right)$$

(4.42)

where h_H is the hub height, κ is the von Karman constant (about 0.4) and z_0 is the ambient
roughness between turbines. From Equation (4.42) we know that the effective roughness z_{00}
increases with C_T but decreases with turbine spacing, as would be expected.

Following the logarithmic profile theory, the impact on hub height wind speed within the
array is then given as

$$\frac{U'_H}{U_H} = \left(\frac{z_{00}}{z_0}\right)^{0.07}\frac{\ln(h_H/z_{00})}{\ln(h_H/z_0)}$$

(4.43)

where U'_H and U_H are the hub height speeds deep within the array and far upstream.

The upper IBL for each turbine is initialized at the height of the top of the turbine rotor,
while the lower one is initialized at the bottom of the rotor. Both IBLs grow with distance

downstream according to the equation

$$h \left(\ln \frac{h}{z_0} - 1 \right) = \left(\frac{x}{z_0} - 1 \right) z_0 \qquad (4.44)$$

where x is the distance from where the IBL is created and z_0 is the downstream roughness (for the above IBL, the turbine roughness, and for the bottom one, the ambient roughness).

The modified wind speed U'_H (decreased) is used as the input of the free-stream wind speed in standard wake models. In the Brower case, the Ainslie model implemented in openWind software was used. A comparison research with five other models has also shown good agreement with measurements [22]. After testing the method with two offshore and two onshore projects, they recommend a turbine roughness value of 1.14 m for offshore sites and 1.30 m for onshore sites [32]. The results have been integrated into openWind, named as the deep array wake model (DAWM).

4.3.4.3   GH WindFarmer model

A slightly different approach has been proposed by Schlez and Neubert (2009) [33]. They used Equation (2.4) in Chapter 2 to determine the height of the IBL, h. An offset of 2/3 hub height is applied to h in order to take into account the momentum not extracted at the ground level, leading to a new height, h'. It is assumed that the disturbance to the ambient wind speed felt at the lower edge of the rotor, z, is decisive for the turbine performance. The wind speed in deep array at the hub height is then calculated following a segmented expression ($z \geq 0.3h'$, $0.09h' < z < 0.3h'$ and $z \leq 0.09h'$).

A geometric model is also included in the GH WindFarmer wake model to take into account the influence of turbine distribution and density on the roughness. Similar to openWind DAWM, WindFarmer also implemented the Ainslie model as the standard single wake model.

The model has been validated with success for two offshore wind farms in the report [33]. The standard wake model starts to underestimate the wind speed deficit from the fourth row and the gap becomes significant from the fifth row. The GH WindFarmer model, however, has shown good agreement with measurement.

Schlez and Neubert (2009) [33] do not validate the method with onshore wind farms in the report. Johnson et al. (2009) [35] have tested the model with a selected large onshore wind farm in North America and confirmed the improvements of deep array wake prediction results for onshore wind farms. They argue that the settings for onshore wind farms must be different from offshore wind farms. For the available data, the best results were achieved with a change in surface roughness length, $z_{02} = z_{01} + 0.03$ m.

4.3.5   Wake Effects in Complex Terrain

The effect of topography has an influence on wake geometry, especially when the terrain is complex, including the narrowing of wind rise and the decrease of the Weibull k factor. Existing engineering-type models for wakes have been developed and calibrated for flat terrain applications [36]. This is a very complicated flow combination requiring advanced numerical flow models to interpret.

Prospathopoulos *et al.* (2008) [36] attempts to simulate the turbine wake effect in complex terrain using a Navier–Stokes solver along with $k-\omega$ turbulence closure. Wind turbines are modelled as momentum absorbers by means of their thrust coefficient. The numerical predictions for one axisymmetric three-dimensional and one quasi-three-dimensional Gaussian hills are compared with those in a flat terrain, presenting some interesting results.

According to Prospathopoulos *et al.* (2008) [36], an increase in turbulence intensity in a complex terrain manifests higher accelerations at the hill top and higher decelerations at the leeside of the hill. This effect is reinforced by the presence of a wind turbine on the hill top. As a result, the velocity deficit remains significant at 20D downstream of the turbine. On the contrary, in the flat terrain case, the deficit has already been practically negligible at 20D. Wake geometry is also modified in a complex terrain. In the flat terrain case, the wake centre is about 0.1D lower than the hub height at 5D downstream, while it is about 0.2D for a complex terrain. The effect of the wind direction on the decay rate of the deficit is also found to be drastic.

4.4 Wind Turbine Micro-siting

Micrositing is the process of strategically placing wind turbines in a wind farm area in order to achieve maximised overall energy production of the wind farm whilst minimising wind turbine loads, and thus is also known as wind turbine layout optimisation. Micro-siting is a meteorological term usually referring to a special scale of less than 2 km.

Adding more turbines to a given site area could be desirable to maximise land use and thereby maximise the overall energy production, but doing so might also lower the average output per turbine because of increased wake loss. There is a point where adding one more turbine may actually lower the overall production of the wind farm. Therefore, the extra investment of adding more turbines has to be justified by the additional revenue in prospect.

Another competing factor entering the game is turbine loads. Wind turbines are usually designed to be in service for 20 years but within the designed limits of wind conditions, including mean wind speed, turbulence intensity, inflow angle and wind shear. Exceeding the limits may cause immature wear and tear of major components, resulting in higher operation and maintenance costs and even earlier retirement of the machine.

Installation costs may also play a role in the decision-making process. It is largely determined by the access roads and electrical collection system connecting wind turbines together as well as the wind turbines themselves. Generally speaking, it would be preferable to keep the required roads and cables shorter if possible and retain the wind turbines in a closer patch. Furthermore, steepness of the terrain, barriers such as rivers and protected areas and the construction of mounting platforms must be considered as well.

Finally, the layout must meet many other regulatory requirements, such as environmental constraints and health concerns (see Chapter 11) and military and civil aviation radar clearance. In a broader sense, micro-siting should also involve understanding the specific wind conditions of the site and thereby selecting the optimal wind turbine type for it. As a result, optimisation of wind turbine layout is not always about finding the best wind speeds or wind resource but is rather a compromise among many balancing goals.

4.4.1   Park Efficiency

Wind park efficiency is the fraction of wind energy extracted from the available wind energy. It is a measure of production loss due to wind speed deficit induced by turbine wake and the prime target for layout optimisation in terms of boosting energy production of the wind park. It is not uncommon to achieve a few percentage of more energy production just by rearranging the layout without increasing the total investment.

The park efficiency, η_{park}, can be expressed as

$$\eta_{park} = \frac{P_{park}}{P_{free}} \times 100\% = 1 - \frac{P_{wake}}{P_{free}} \times 100\% \qquad (4.45)$$

where P_{park} is the predicted energy production of a wind park, P_{free} is the production at the free-stream wind speed, that is the energy available without considering wake effects, and P_{wake} is the production loss caused by wake effects.

Figure 4.9 shows a wind turbine layout of a regular grid where the row is defined as perpendicular to the prevailing wind direction. Each turbine may be exposed to different levels of wake deficit depending on where it is in the wind park, hence presenting different efficiencies. In general, the overall park efficiency should be higher than 93% while that of an individual turbine should be higher than 85%. Otherwise, it may be wise to improve the layout further or even delete some of the least performing turbines. This is a good reason to keep the row spacing at least larger than 5 rotor diameters in the prevailing wind direction to limit production loss due to wake effects and 3 rotor diameters in the crosswind direction to mitigate the risk of wake turbulence for onshore wind farms. For offshore and cold climates where low turbulence intensity and stable stratification prevails, the spacing should be even larger; 7D in windward and 5D in crosswind directions can be taken as the minimum. Extending row spacing can have

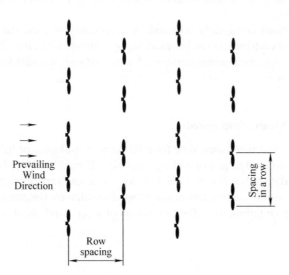

Figure 4.9   Wind turbine layout with a regular grid

a considerably positive effect on park efficiency, more so for stable air conditions. Keeping the number of rows low, within 5 rows, is usually a good idea, especially when the prevailing wind direction is relatively focused.

4.4.2   Capacity Factor

The capacity factor (CF) of wind power is the ratio of actual delivered power to the theoretical maximum possible, that is running full time (24 hours a day 365 days a year) at rated power:

$$CF = \frac{\text{actual power production}}{\text{rated power running full time}} \times 100\% \qquad (4.46)$$

CF can be computed for a single turbine, a wind farm consisting of many turbines or an entire country consisting of hundreds of wind parks. Besides the geographical location, it is also a matter of turbine design. A larger rotor combined with a smaller generator will achieve a higher capacity factor. A higher CF does not indicate higher efficiency, or vice versa, but you can generally say that a higher CF is better and, in particular, more economical, because the net present return of a wind turbine is proportional to its average CF over the 20 years lifetime. Consequently, the CF value is crucial for decision makers. The typical capacity factors for wind parks are 20~40%.

4.4.3   Site-Specific Wind Conditions

The correct information of wind conditions of the wind project is a prerequisite for micro-siting. The information is important because in part the wind turbine in question will have to operate within its designed limits, which have to be closely monitored during the micro-siting process.

The wind conditions are usually provided by a summary of wind data measured on-site. Wind conditions for each turbine can be simulated by software tools, introduced in the previous chapter. The most influential wind conditions for the purpose of micro-siting are given in the following subsections.

4.4.3.1   Annual Mean Wind Speed

The annual mean wind speed should be for a full year or multiples of full years. Partial year data should never be used so as to avoid seasonal bias. If possible, it is always a good idea to use multiple year data rather than just one full year as annual variations in mean wind speeds can be significant. A preliminary layout can be used to calculate the annual mean wind speed at hub height for each turbine site. The calculation of mean wind speed is to be introduced in Chapter 5.

4.4.3.2   Wind Rose

The prevailing wind direction provides a guideline for turbine layout as stated previously. The tool describing the wind directional distribution is the polar plot, called a wind rose (see Figure 4.10). There are two types of wind rose: the frequency rose and the energy rose.

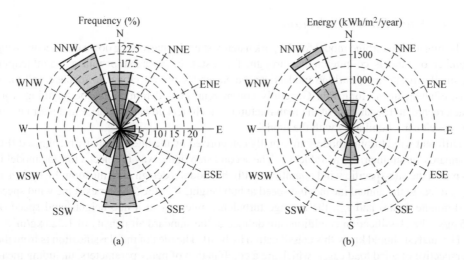

Figure 4.10   Wind frequency rose (left) and energy rose (right) for the same set of wind data created with WindPRO 2.7

The frequency rose indicates the frequency of occurrence by direction and the percentage of time the wind blows in certain speed ranges in adirection bin, or sector, and can be displayed by segments of bands in different colours. The wind energy rose indicates wind energy coming from each direction bin in kW h/m²/year (absolute) or as a percentage of the total energy (relative) and is a clearer representation of the prevailing wind direction as it indicates the energy content of the wind.

The wind roses are plotted by sorting wind data into a desirable number of direction bins, or sectors, typically 12 or 16, and then calculating the relevant statistics for each bin.

The frequency rose and the absolute energy rose can be computed as

$$f_i = \frac{N_i}{N} \times 100 \qquad (4.47)$$

$$E_i = \sum_j P_{ij} \qquad (4.48)$$

where N_i refers to the number of records in the direction sector i, N is the total number of records in the data set and P_{ij} is the power density for the record j in the direction sector i.

4.4.3.3   Turbulence Intensity

The turbulence intensity, I_{ref}, at 15 m/s provides a reference for preliminarily checking the suitability of a turbine model. Engineers working for the manufacturer may require a more detailed description of turbulence intensity, such as the frequency distribution of turbulence, average and extreme turbulence intensities for different wind speeds in the power curve and wake-added turbulence in order to provide warranty for the machine. It is very likely to change the layout should turbulence intensity be found exceeding the designed limits. Some sites could be eliminated entirely for these reasons.

Other worth-knowing wind conditions include inflow angle, wind shear across the rotor, air density, 50-year extreme wind speed and gust.

4.4.4   Wind Turbine Selection

Selecting a wind turbine model for the park usually starts by narrowing down the ever-growing number of models commercially, through, for instance, partnership with a manufacturer. A smaller portfolio of turbine models can be easier for the developer to manage and more cost-efficient in the long run. Generally commercial considerations can be anything from track record and reputation of the manufacturer to technical advantage, price and warranty, to the developer's familiarity with the models.

Different wind turbine models are usually categorised by the internationally recognised IEC standards, as summarised in Table 4.1. The second step in the way to select a turbine model is to match the site-specific wind conditions with the model's design classification. In Table 4.1, V_{ave} refers to the annual mean wind speed at hub height, V_{ref}, the 50-year extreme wind speed (10-minute mean) and I_{ref}, the average turbulence intensity at a hub-height wind speed of 15 m/s. The classification conditions are defined at the standard air density of 1.225 kg/m^3.

The analyst should know this classification by heart. The idea of the classification is from the perspective of wind load cases, which are a combination of many parameters, including mean wind speed distribution, air density, turbulence intensity, inflow angle, wind shear and extreme wind conditions. A different value on one of those parameters may allow an adjustment to the others, but it has to be made in consultation with the manufacturer.

The classification is composed of two parts: mean wind speed and turbulence intensity. For instance, IEC IA turbines are designed to withstand an annual mean wind speed of up to 10 m/s and a maximum mean turbulence intensity at 15 m/s of 16%, the greatest wind loads in the classification. The turbines are usually made stronger at the expense of lower capacity factors. It can be done by using smaller rotor diameters relative to rated power or by changing pitching strategy of the blades to lower the angle of attack, thereby absorbing less energy from the wind (other solutions are possible, such as a dual generator). Therefore, two things could happen if the turbine were not suitable for the site conditions: (1) producing less energy than could have been the case if higher-class turbines had been used for lower wind conditions, making them less cost-efficient (lower capacity factor), or (2) immature wear and tear of the machine for the opposite situation, that is the turbines were not designed to bear the wind loads (a higher capacity factor with much higher operating costs).

Wind conditions of different sites within a wind park may be significantly different, leading to more than one suitable wind class. In this case, the classification of the wind park is usually defined by the windiest and most turbulent sites and other sites with lower wind conditions should follow suit. It is possible, however, to employ two different turbine models or two

Table 4.1   Wind turbine classification derived from IEC 61400-1 (2005) for the standard air density of 1.225 kg/m^3 [37]

Wind turbine class		I	II	III	S
V_{ave} (m/s)		10	8.5	7.5	
V_{ref} (m/s)		50	42.5	37.5	
$V_{50,gust}$ (m/s)		70	59.5	52.5	User defined
I_{ref}	A	0.16			
	B	0.14			
	C	0.12			

hub heights for the same model, each corresponding to a different IEC class in order to fully exploit variations in the wind resource, but it is only recommended when the turbine models are similar and from the same manufacturer.

Extreme wind conditions may also determine the wind class. In the IEC class, the 50-year extreme mean wind speed is taken as 5 times the annual mean wind speed, but in reality wind conditions may be outside this envelope. In coastal Southern China, for example, where typhoons prevail in the summer, V_{ref} usually points to IEC I, but the V_{ave} usually falls in the category of IEC II and III. IEC I turbines have to be employed in this case if no user-defined models designed for such a situation are available.

Once the shortlist of turbine models is made, a layout can be intended for each turbine model based on its rotor diameter and designed hub height. Production for each solution is then estimated and compared with the corresponding capital and operating costs before making the final decision. The commercial department will usually take over from this point.

4.4.5   Site Survey

Because the main objective of micro-siting is to quantify the small-scale variability of the wind resource over the terrain of interest, a close-up observation through a site survey is crucial. Published work relating to a wind resource assessment usually refers a site survey to a wind monitoring campaign, but the scope here is much wider.

A site survey should work hand in hand with computer modelling in order to achieve the optimal micro-siting solution and a more accurate production estimate with minimised uncertainties (see Figure 4.11). Computer modelling makes qualitative judgements of a micro-scale wind resource for each proposed turbine site, while a site survey provides intuitive information on whether the wind resource modelling software in use may or may not violate its operational constraints and if anything locally in particular may affect the modelling results. For instance, a digital height contour map may not represent the true terrain very well; there may be things that wind turbines have to stay away from such as power lines, communication towers, community concerns, etc. As a result, the layout will most likely have to be adjusted during the site survey.

4.4.5.1   Preparing for a Site Survey

The site survey may involve intensive travelling and hard work in the wild as each and every turbine spot should be visited; one would not want to attempt it without good preparation beforehand. Before a site survey, the engineer should have gained a good picture of the wind climate, an optimised layout based on the best knowledge available and particularities of each turbine site that need to be checked out on-site.

A global positioning system (GPS) handset is the most critical tool to carry. It can be used to mark coordinates and tracks, measure distance and orientation, and provide information on altitude, maps, moving speed, time of sunset, etc. It can also help us take directions and find our way in places with few human traces. Coordinates of the preliminary layout and met mast locations should be imported into the GPS device prior to the trip. Backup batteries are strongly recommended. Other tools that can be helpful include a compass, telescope with range metering, local map and an inclinometer.

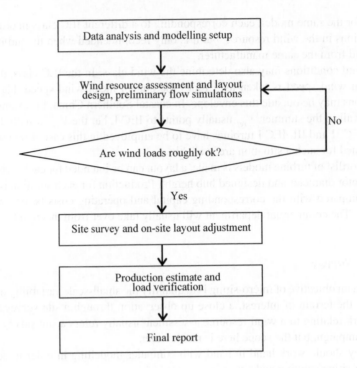

Figure 4.11   Typical micro-siting procedure

In some cases, the job may be done by driving a car around most of the time, but it is not uncommon to have to hike all day for a few days in the wild, especially in complex terrain. The engineer should be well prepared both physically and mentally. Basic wild survival skills and experience would be very helpful. For safety reasons, one should not attempt a site survey like this alone without at least one companion and never stay overnight in the wild. In the backpack, there should be enough drinking water, preferably sports drinks providing electrolytes, food that is high in carbohydrates, a flashlight, knife and whistle for sending signals in case of an emergency. An investigation on the type of harmful wildlife on the site in advance, such as insects and snakes, is also a good idea, so that you are able to take preventive action. Summer time in tropical areas and snow-covered winter should be avoided altogether for an intensive site survey. Figure 4.12 was taken during a difficult site survey in Southern China, giving a rough impression of the working situation in the wild.

4.4.5.2   Missions to Accomplish

Verifying the Met Masks
The verification of met masks mainly consists of checking coordinate discrepancies, met mast arrangements and the surroundings. This job becomes extremely important when the wind data and documents are provided by other parties and the engineer has doubts on certain aspects of the data.

Figure 4.12   Site survey for a wind park in Southern China

The coordinates presented in documents and wind data may not be accurate and a small bias of the coordinates may lead to a significantly different result of the wind resource assessment, especially for a complex terrain. Thus, they have to be verified with a GPS on-site.

The arrangements of met masts have to meet stringent requirements in order to minimise measurement uncertainties and biases caused by the tower and supporting structures and accessories (see Chapter 8). A good understanding of the met masts will help us analyse wind data, identify suspicious phenomena and most importantly assess the uncertainties of the wind data.

The surroundings are also important for wind data analysis as they may help make sense of some phenomena in the data, such as high turbulence intensity in certain directions. There could be some topographic features including obstacles and wind breaks that need to be included or emphasised in wind flow modelling as well.

Investigating Surface Roughness and Wind-Obstructing Obstacles

The surface roughness based on satellite images may be very different from on-site observations. Furthermore, further development of roughness elements such as growing trees and urban development should be taken into consideration for assessing energy production of the wind park for the next 20 years.

The shelter effects of obstacles have become somewhat less important in wind flow modelling in recent years as utility scale wind turbines keep growing in height and are being built further away from urban structures. However, they should be investigated and mapped for each turbine site nevertheless.

Fine Adjustments of the Layout

Many local features can only be scrutinised on-site. For instance, the turbine site may be slightly off the crest, the slopes may be too steep, there could be overhead power or communication lines and height variations may be more severe than imagined. Therefore, each turbine site has to be visited and adjusted according to on-site observations. Adjusting one position may lead to adjusting an array of them.

Apart from avoiding obvious objections and concerns, the prime goal of an on-site layout adjustment is to identify features that may lead to exceeding wind loads and to make an adjustment accordingly. Obstacle shelter effects introduced in Chapter 2 can be the handiest analytical tool for such a purpose (e.g. see Figure 2.21).

Taking Pictures and Notes for Reports
Panoramic views of the park from a few critical points, shots of representative roughness elements and influential features and photos of met mast arrangements are important to describe the wind park vividly in reports. Any adjustments made and influential features identified should be carefully noted for later use. The engineer should make sure that the lens direction and location of the pictures taken can be identified afterwards.

4.4.6 Wind Sector Management

Wind sector management is a wind turbine control strategy used to optimise the performance of the wind park as a whole or to lower loads for certain wind turbines in the park for safety reasons by de-rating or shutting down the turbine in certain direction sectors, as illustrated in Figure 4.13.

It is not always optimal to run the turbines in the first row at full rating in certain direction sectors, because the wake-induced velocity deficit for downstream turbines may outweigh the production increased by the first row. De-rating the wind turbines in the first couple of rows or in patterns (e.g. every second turbine) in these wind directions may lead to a more balanced power production and turbine loads across the wind park. The total power production may also increase [38]. With this control strategy, it is possible to propose a layout with larger spacings in the prevailing wind direction in order to significantly increase park efficiency and smaller than three-dimensional spacings in the crosswind direction to increase land use while keeping the turbines safe, that is de-rating the turbines in the crosswind direction.

In complex terrain, exceeding wind loads at some spots, such as turbulence intensity and inflow angle, may only come from certain wind directions with insignificant wind resource. These spots would probably have to be given up without wind sector management to shut down the turbines in these dangerous directions. Therefore, a good wind sector management strategy

Figure 4.13   Wind sector management

may lead to a more efficient land use and thereby higher installed capacity in complex terrain. Wind sector management in these scenarios will decrease energy production of the individual turbines and consequently lowering their capacity factors. The production loss caused by wind sector management has to be evaluated in order to justify the decision.

4.5 Summary

This chapter started with introducing wind energy density and wind energy conversion coefficients, C_p, C_T and C_e. C_p and C_T denote the fraction of energy and momentum extracted from the wind by a turbine respectively and C_e is the fraction of wind energy actually converted into electrical power. C_T is one of the most critical parameters in wind turbine wake models as it is derived from the conservation of momentum.

Velocity deficit and added turbulence in wake are usually modelled separately even though turbulence itself has important impacts on the development of the wake structure. The wake can be analytically divided into three zones: the near wake, the intermediate wake and the far wake. The near wake is usually 2 to 3 rotor diameters in length immediately downstream of the turbine rotor and is very turbulent due to intensive mixing induced by the pressure gradient. Most single wake models including those of Jensen and Ainslie ignore the complexity of the near wake and therefore become invalid in the near wake zone. The ending of intermediate wake is marked when the velocity in the centreline starts to recover, extending around 3 rotor diameters behind the near wake. It is recommended that for onshore conditions the minimum spacing between two neighbouring turbines in the prevailing wind direction should be larger than 5D and 3D in the crosswind direction for turbine safety.

Atmospheric stability and turbulence intensity affects mixing of the wake with the adjacent free-stream wind, thereby changing the ability of momentum recovery of the wake. This explains why the wake is usually longer and more pronounced offshore and in cold climate where stable air stratification and low turbulence intensity are present. Thus, turbine spacings for offshore wind parks should be larger than onshore ones so as to achieve the same park efficiency.

Standard wake models do not take into account the dynamic feedbacks of the turbines to the atmospheric boundary layer, assuming a one-way interaction between the turbines and the boundary layer. This is not true as the wind park becomes sufficiently large, especially for offshore conditions. Therefore, the standard wake models tend to underestimate the velocity deficit in a deep array. Two validated deep array models implemented by openWind and WindFarmer have been introduced.

With a good understanding of wind turbine wakes, it is then possible to talk about wind turbine layout optimisation or micro-siting, which is an effort to maximise the energy output of a wind park while keeping turbine loads under control.

References

[1] Betz, A. (1966) *Introduction to the Theory of Flow Machines* (D.G. Randall, Trans.), Pergamon Press, Oxford.
[2] IEC 61400-12-1 (2005) Wind Turbines – Part 12-1: Power Performance Measurements of Electricity Producing Wind Turbines.
[3] Muljadi, E., Butterfield, C.P. (1999) Pitched-Controlled Variable-Speed Wind Turbine Generation. Presented at the 1999 *IEEE Industry Application Society Annual Meeting*, Phoenix, Arizona.

[4] Svenningsen, L. (2010) *Power Curve Air Density Correction and Other Power Curve Option in WindPRO*, EMD International A/S, Denmark.

[5] Vermeer, L.J., Sørensen J.N., Crespo, A. (2003) Wind Turbine Wake Aerodynamics. *Progress in Aerospace Sciences*, **39**, 467–510.

[6] Méchali, M., Jensen, L.E., Barthelmie, R.J., Frandsen, S.T., Réthoré, P. E. (2006) *Wake Effects at Horn's Rev and Their Influence on Energy Production. EWEC*, Athens.

[7] Natalia, M., Krzysztof, R. (2010) Study of Wake Effects for Offshore Wind Farm Planning. *Modern Electric Power System*, Wroclaw, Poland.

[8] Werle, M.J. (2008) A New Analytical Model for Wind Turbine Wakes. Report No. FD 200801, FloDesign Inc., Wilbraham, USA.

[9] Manwell, J.F., McGowan, J.G., Rogers, A.L. (2002) *Wind Energy Explained, Theory, Design, and Application*. John Wiley & Sons, Ltd, West Sussex.

[10] Barthelmie, R.J., Folkerts, L., Larsen, G.C., Rados, K., Pryor, S.C., Frandsen, S.T., Lange, B., Schepers, G. (2005) Comparison of Wake Model Simulations with Offshore Wind Turbine Wake Profiles Measured by Sodar. *Journal of Atmospheric and Oceanic Technology*, **23**, 888–900.

[11] Douwe, J.R. (2007) Validation of Wind Turbine Wake Models Using Wind Farm Data and Wind Tunnel Measurements. MSc Thesis, Faculty of Aerospace Engineering, Delft University of Technology, The Netherlands.

[12] Jensen, N.O. (1983) A Note on Wind Generator Interaction. Risø National Laboratory, Risø-M-2411, Denmark, 16pp.

[13] http://www.wasp.dk/ (last accessed on 30 December 2013).

[14] http://www.emd.dk/ (last accessed on 30 December2013).

[15] http://www.gl-garradhassan.com/en/software/GHWindFarmer.php (last accessed on 30 December 2013).

[16] Sørensen, T., *et al.* (2008) *Adapting and Calibration of Existing Wake Models to Meet the conditions inside offshore wind farms*, EMD International A/S, Denmark.

[17] Nielsen, P., *et al.* (2006) *The WindPRO Manual*, edition 2.5, EMD International A/S, Denmark.

[18] Rudion, K. (2008) Aggregated Modelling of Wind Farms. Otto-von-Guericke-Universität, Magdeburg.

[19] Larsen, G.C., *et al.* (2003) Mean wake deficit in the near field. *European Wind Energy Conference and Exhibition 2003*, Madrid, Spain.

[20] Ainslie, J.F. (1988) Calculating the flow field in the wake of wind turbines. *Journal of Wind Engineering and Indudtrial Aerodynamics*, **27**, 213–224.

[21] Lange, B., Waldl, H.P., Guerrero, A.G., Heinemann, D., Barthelmie, R.J. (2003) Modelling of Offshore Wind Turbine Wakes with the Wind Farm Program FLaP. *Wind Energy*, **6**, 87–104.

[22] Philippe, B., *et al.* (2012) Overview of Six Commercial and Research Wake Models for Large Offshore Wind Farms. *Proceedings of EWEA 2012*.

[23] Crespo, A., Hernández, J., Fraga, E., Andreu, C. (1988) Experimental Validation of the UPM Computer Code to Calculate Wind Turbine Wakes and Comparison with Other Models. *Journal of Wind Engineering and Industrial Aerodynamics*, **27**, 77–88.

[24] Rafael, G.M., Crespo A., Migoyab, E., Manuelb, F., Hernándezc, J. (2005) Anisotropy of turbulence in wind turbine wakes. *Journal of Wind Engineering and Industrial Aerodynamics*, **93**, 797–814.

[25] Schepers, G., Barthelmie, R., Rados, K., Lange, B., Schlez, W. (2001) Large Off-shore Windfarms: Linking Wake Models with Atmospheric Boundary Layer Models. *Wind Engineering*, **25**(5), 307–316.

[26] Frandsen, S.T. (2007) Turbulence and Turbulence-Generated Structural Loading on Wind Turbine Clusters. Risø-R-1188 (EN), Risø National Laboratory, Denmark. Available at: http://orbit.dtu.dk/files/12674798/ris_r_1188.pdf (last accessed on 15 August 2014)

[27] Frandsen, S.T., *et al.* (1999) *Model for Turbulence in Wind Farms*, Risø National Laboratory, Denmark.

[28] Dekker, J.W.M., Pierik, J.T.G. (1998) *European Wind Turbine Standards II*, ECN Solar & Wind Energy, The Netherlands.

[29] Quarton, D.C., Ainslie, J.F. (1989) Turbulence in Wind Turbine Wakes. Presented at the *EWEC 1989 Conference*.

[30] Crespo, A., Hernández, J. (1996) Turbulence Characteristics in Wind-Turbine Wakes. *Journal of Wind Engineering and Industrial Aerodynamics*, **61**, 71–85.

[31] Rathmann, O., Rebecca, B., Frandsen, S. (2006) Turbine Wake Model for Wind Resource Software. *European Wind Energy Conference 2006, Scientific Proceedings*.

[32] Brower, M., Robinson, N. (2012) *The openWind Deep-Array Wake Model: Development and Validation*, AWS Truepower, USA.

[33] Schlez, W., Neubert, A. (2009) New Developments in Large Wind Farm Modeling. *Proceedings from the EWEA Conference 2009*, Marseille, France, 8 pp.

[34] Frandsen, S., Barthelmie, R.J., Pryor, S.C., Rathmann, O., Larsen, S.E., Højstrup, J., Thøgersen, M. (2006) Analytical Modelling of Wind Speed Deficit in Large Offshore Wind Farms. *Wind Energy*, **9**, 39–53.

[35] Johnson, C., Graves, A.M., Tindal, A., Cox, S., Schlez, W., Neubert, A. (2009) New Development in Wake Models for Large Wind Farms. Poster Presentation at the 2009 *AWEA Windpower Conference*, Chicago, USA.

[36] Prospathopoulos, J.M., Politis, E.S., Chaviaropoulos, P.K. (2008) Modelling Wind Turbine Wakes in Complex Terrain. Centre of Renewable Energy Sources, Wind Energy Department, EWEC, Greece.

[37] IEC 61400-1 (Ed.3) (2005) Wind Turbine – Part 1: Design Requirements [S].

[38] Carlén, I. (2008) *Some Simple Concepts for Dynamic Wind Sector Management of Wind Farms*, Teknikgruppen AB, Sweden.

5

Wind Statistics

It is a matter of common observation that the wind is not steady and varies constantly with time and space. Given its randomness, it is necessary to study the wind statistically, that is the probability density distribution of the wind speeds. In the language of statistics, it is unpredictable in the short run but has a regular and predictable pattern in the long run. Statistical analysis is thus an indispensable part of the wind resource assessment and wind data analysis.

This chapter will focus on the statistical study of wind speed. For those unfamiliar with statistics, a brief review of important concepts will be given as well, as one of the purposes in this chapter is to understand the language of statistics and probability without going into complicated mathematical textbooks on statistic study.

Cup anemometers, which only measure the horizontal component of the wind speed, are assumed in this chapter, coinciding with most published wind energy standards.

5.1 Statistics Concepts Review

This section provides a brief review of concepts in statistics that are important to understand the wind. For a more detailed introduction in statistics, readers are referred to numerous monographs, for instance Hwei (2010) [1] and Wilks (1962) [2].

5.1.1 Random Variables

A random variable is a variable whose possible values are numerical outcomes of a random phenomenon. A discrete random variable is one that may take on only a countable number of distinct values while a continuous random variable takes on an infinite number of possible values. Measurements, such as wind speed, temperature and air density, are continuous random variables as they vary continuously.

A random variable seems to be stochastic, but it usually has a regular and predictable pattern in the long run. Tossing a coin, for instance, may give one of the two values (heads or tails) randomly each time, but the outcomes of many coin tosses will show a predictable pattern of 50% probability for either value to occur. Therefore, a random phenomenon is a situation in

Wind Resource Assessment and Micro-siting, Science and Engineering, First Edition. Matthew Huaiquan Zhang.
© 2015 John Wiley & Sons, Ltd. Published 2015 by John Wiley & Sons, Ltd.

which we know what outcomes could happen, but we do not know which particular outcome did or will happen.

Wind statistics is a science that describes the patterns of the wind speed over representative time periods as a continuous random variable.

5.1.2   Sample Mean and Standard Deviation

The mean of a random variable is the sum of every possible value weighted by the probability of that value, that is its weighted average. The sample mean of a group of observations gives each observation equal weight and can be expressed as

$$\mu = E[x] = \frac{1}{n}\sum_{i=1}^{n} x_i \tag{5.1}$$

Here the operator E denotes the mean or expected value of x.

The sample standard deviation is the square root of the variance that measures the spread or variability of the samples of the random variable. It is given as

$$\sigma = \sqrt{\frac{1}{n-1}\sum_{i=1}^{n}(x_i - \mu)^2} \tag{5.2}$$

or

$$\sigma = \sqrt{E[x - \mu]^2} = \sqrt{E[x^2] - (E[x])^2} \tag{5.3}$$

In statistics and probability theory, the standard deviation shows how much variation or dispersion from the average exists. A low standard deviation indicates that the data points tend to be very close to the mean (also called the expected value); a high standard deviation indicates that the data points are spread out over a large range of values. The standard deviation is also commonly used to measure confidence in statistical conclusions.

5.1.3   Probability Density Distribution

Probability of any outcome of a random phenomenon is the proportion of times the outcome would occur in a very long series of repetitions. The probability of observing any single value is equal to zero (which does not mean it will not happen), since the number of values that may be assumed by the random variable is infinite. Any probability is a number between 0 and 1. The total probability of all possible outcomes must be equal to 1. The probability that an event does not occur is 1 minus the probability that the event does occur. For two independent events (with no outcomes in common), the probability for either one to occur is the sum of their individual probabilities [2].

A probability density function (PDF) of a continuous random variable describes the relative likelihood for this random variable to take on a given value. The probability density function is non-negative everywhere, and its integral over the entire space is equal to one. The probability of the continuous random variable falling within a particular range of values is given by the area under the density function curve but above the horizontal axis and between the lowest and greatest values of the range, that is the integral of the density function over that range.

A cumulative distribution function (CDF), or just a distribution function of a continuous random variable x, describes the probability that X will be found at a value less than or equal to a given value X, that is the integral of the density function over the range from minus infinity to x. The CDF can be expressed as

$$P = F(x) = \int_{-\infty}^{X} f(x)dx \tag{5.4}$$

where $f(X)$ is the PDF of a continuous random variable X.

The mean (also called the expected value) of variable X is computed by taking the product of each possible value x of X and its probability $f(x)$ and can be expressed as

$$\mu = E[x] = \int_{-\infty}^{+\infty} xf(x)dx \tag{5.5}$$

The standard deviation of X with the probability density function $f(x)$ is

$$\sigma = \sqrt{\int_{-\infty}^{+\infty} (x - \mu)^2 f(x)dx} \tag{5.6}$$

The nth central moment of X is given by

$$\mu_n = \int_{-\infty}^{+\infty} (x - \mu)^n f(x)dx \tag{5.7}$$

The skewness is defined as the normalized third central moment, which measures the asymmetry of the probability distribution of a real-valued random variable about its mean. A symmetric PDF will have a skewness of zero. A distribution skewed to the left (the left tail is longer; the mass of the distribution is concentrated on the right of the figure) will have a negative skewness. Conversely, a distribution skewed to the right will have a positive skewness. It can be given as

$$\gamma = \frac{\mu_3}{\sigma^3} \tag{5.8}$$

The kurtosis is defined to be the normalised fourth central moment. It measures the 'peakness' of the PDF of a real-valued random variable, that is whether the distribution is tall and skinny or short and squat, compared to the normal distribution of the same variance. The kurtosis is usually taken as the normalised fourth central moment minus 3. If a distribution has a higher peak than the normal distribution at the mean and longer tails, the kurtosis will be positive; conversely, bounded distributions tend to have low kurtosis. It can be given as

$$k = \frac{\mu_4}{\sigma^4} - 3 \tag{5.9}$$

Skewness and kurtosis offer a way to characterise the shape of an unknown probability distribution.

5.2   Wind Data Time Series

Samples have to be collected as references in order to statistically describe the wind. This is usually done by sampling the wind every few seconds and then summarising the samples over a certain time period (usually 10 minutes or 1 hour) into sample means, standard deviations, maximum and minimum values. Usually only the summarised time-step data series are stored in a data logger as the raw wind data for wind resource assessment and the details within the each time period are ignored. Before computing wind statistics, time series have to be checked for homogeneity, such as the replacement of instruments.

5.2.1 Mean Wind Speed

Suppose that the wind speed is sampled every 2 seconds and summarised into a mean and standard deviation over every 10 minutes (averaging time period). Figure 5.1 shows the 300 data points recorded over one of the 10-minute periods; the mean wind speed value is indicated by the dashed line in the centre.

Following Equations (5.1) and (5.2), we can calculate the sample mean wind speed and standard deviation as

$$\mu_v = \frac{1}{300} \sum_{i=1}^{300} v_i \tag{5.10}$$

$$\sigma_v = \sqrt{\frac{\sum_{i=1}^{300} (v_i - \mu_v)^2}{299}} \tag{5.11}$$

The product is a 10-minute averaging mean wind speed and standard deviation time-sequence data series. The so-produced raw wind data are a series of discrete random values, which are further deduced from a finer set of discrete random values rather than the sampled transient data. The randomness of mean wind speeds in the raw data is more tamed than the sampled data points because much has been cancelled out during the averaging calculations. The randomness or 'gustiness' within the 10-minite averaging time is represented by the standard deviation, which is usually studied separately as turbulence

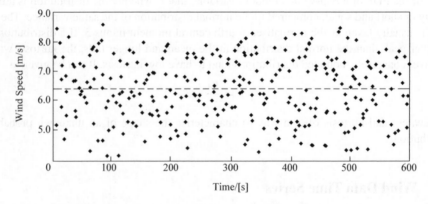

Figure 5.1 Transient wind speed data points recorded every 2 seconds for 10 minutes (600 s). Dashed line in the centre represents the mean wind speed value over this period

intensity. This idea corresponds to Equation (3.6) in that a meteorological variable can be studied as a mean plus a superimposed fluctuating term.

5.2.2 Turbulence Intensity

Turbulence is very harmful for wind turbine components because the fatigue loads of a number of major components in a wind turbine are mainly caused by turbulence. Turbulence impacts wind energy in several other ways as well, specifically through power performance effects, wake effects and noise propagation [3]. Therefore, the knowledge of the turbulence intensity of a turbine site is of vital importance.

Individual turbulent motions are essentially unpredictable, but some properties of turbulence can be effectively analysed in a statistical sense [3]. In the wind energy industry, turbulence intensity, I, is used as a metric of turbulence to quantify how much the wind varies typically within 10 minutes. It is the standard deviation of the horizontal wind speed divided by the mean wind speed over the same time period (normalised standard deviation), typically 10 minutes:

$$I = \frac{\sigma_v}{\mu_v} \tag{5.12}$$

As Equation (5.12) indicates, for a certain mean wind speed, rapid fluctuation of the wind or wider spreads of the sample wind speed data will give high turbulence intensity. Conversely, steady winds will represent a lower turbulence intensity value.

Turbulence Intensity Curve

Normally, the standard deviation of wind speed at a given location increases more slowly than the mean wind speed, indicated by a downwards curve shown in Figure 5.2. In other words, a higher wind speed at a given location is usually accompanied by a smaller turbulence intensity. IEC 61400-1 (2005) [4] uses the turbulence intensity at 15 m/s mean wind speed as the reference for wind turbine classification.

If the turbulence intensity curve in Figure 5.2 tends to tilt upwards at the higher wind speed end, this fact should be treated with caution because a combination of high wind speed and high turbulence intensity will dramatically increase turbine loads. This phenomenon could happen in a complex terrain or offshore. For offshore, a higher wind speed usually stirs up bigger waves, which may in turn generate higher turbulence intensity in the wind.

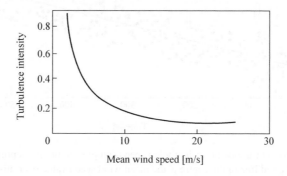

Figure 5.2   Generalised turbulence intensity and mean wind speed curve

Trended Turbulence Intensity

The turbulence intensity calculated from Equation (5.12) considers turbulence as a stationary stochastic process imposed on a given constant mean wind speed. This is not true when the mean wind speed changes slowly with time, displaying a trend in the wind speed time series, as shown in Figure 5.3. Often, a linear trend is assumed [5]. The trending effect in the wind speed time series causes overestimated standard deviations, resulting in an erroneous high turbulence, because the standard turbulence intensity calculation does not distinguish between these trends and the 'real' turbulence.

It is not possible to directly calculate the trend contribution from the typical statistics of a 10-minite mean and standard deviation. Hansen and Larsen (2006) [5] has proposed a linear algorithm for a de-trending turbulence standard deviation based on time series statistics only and reported that the de-trending turbulence intensities are reduced in the range of 3–15% compared to the raw turbulence intensity and 6–8% for offshore sites for the whole operational range for a wind turbine. Damaschke *et al.* (2007) [6] compares a few de-trending methods.

De-trending turbulence intensity is particularly important for wind turbine load simulation. The trending turbulence intensity is false, partly because wind turbines should be able to respond to the slow wind speed changes within seconds by pitching blades, for example, and turbulence would act in a different way on wind turbines.

The trending effect falsely increases turbulence intensity dramatically during the frontal passage of weather systems. The passage of weather fronts can usually be indicated by sudden changes in wind direction and air temperature. Unrealistic turbulence spikes of this kind may not be significant for calculating average turbulence intensity of a given mean wind speed because the large pool of data points should have swallowed the effect of a few extreme values, but they have to be excluded when it comes to extrapolating extreme turbulence intensity to avoid severe errors.

Averaging Time

Apart from trends in wind speed, averaging time also plays an important role in the calculation of turbulence intensity. Turbulent motions are of all sizes with a large range of temporal and

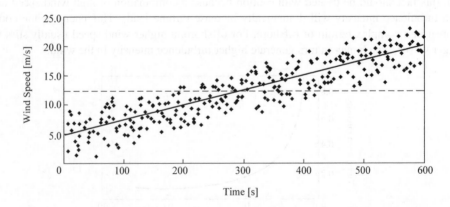

Figure 5.3   Illustration of trended transient wind speed samples. Solid line represents the linear trend of the sample data; dashed line in the centre is the mean wind speed value over this period

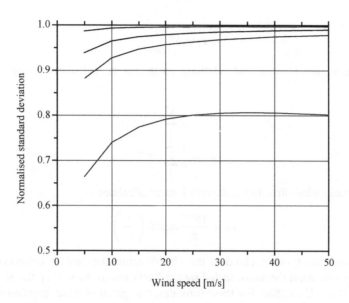

Figure 5.4 Standard deviations of different wind speeds and averaging times (from the top: 30 min, 10 min, 5 min and 1 min), all normalised relative to a 60-minute averaging time for neutral stability

spatial scales. It is stated in IEC 61400-12-1 that the averaging time for large utility scale wind turbines should be 10 minutes, coinciding with the averaging period for most contemporary meteorological data, but for turbines with a rotor diameter smaller than 16 m, the averaging time should be reduced to 1 minute [7]. This is because the turbulence size of interest should be of a similar scale to that of the turbine rotor. The averaging calculation basically treats averaging time as the benchmark between the mean and the superimposed turbulence term.

Statistically, longer averaging times will generally give larger turbulence intensities for the range of averaging times relevant for measurements of atmospheric turbulence intensity (1–60 minutes) and the discrepancy is greater at lower wind speeds, as shown in Figure 5.4. The variation of turbulence intensity with averaging time and wind speed can be investigated by integrating models of the spectral distribution of wind speed fluctuations. The Kaimal spectra model [8] (see Section 9.6.3) for neutral stability has been applied to produce Figure 5.4.

To access the available wind energy at a certain site more thoroughly, high-frequency sampling of the wind speed using lidar and sodar may be necessary.

5.2.3 Wind Direction

The wind direction is a circular function with values between 1 and 360 degrees. It becomes problematic when linear techniques are applied to circular data. Therefore the wind direction discontinuity at the beginning/end of the scale requires special processing to compute a valid mean and standard deviation [9].

US EPA (2000) [9] recommends the Mitsuta methodology, documented in Mori (1986) [10], to compute the mean wind direction, but it can also be carried out on the radius using the transformation:

$$\theta_i = \theta_i^\circ \frac{\pi}{180^\circ} \tag{5.13}$$

Using radians, the Easting and Northing mean can be calculated using trigonometric functions as

$$\mu_x = \frac{1}{n} \sum_i \cos \theta_i \tag{5.14}$$

$$\mu_y = \frac{1}{n} \sum_i \sin \theta_i \tag{5.15}$$

The sample mean wind direction in degrees is then calculated as

$$\mu_\theta = \frac{180^\circ}{\pi} \arctan \left(\frac{\mu_x}{\mu_y} \right) \tag{5.16}$$

Due to the complexity of circular data, the wind direction standard deviation is usually computed by approximation formulas, including the Yamartino method [11], the Mardia method [12] and the Nomad2 method. For these formulas, the spread of wind direction data points is defined as

$$\varepsilon = \sqrt{1 - (\mu_x^2 + \mu_y^2)} \tag{5.17}$$

The effect of ε for the same mean wind direction is shown in Figure 5.5. The value of ε is between 0 and 1; smaller ε means a smaller directional standard deviation.

The Yamartino method [11] is implemented in the NRG Symphonie logger [13], as given in the following:

$$\sigma_\theta = \frac{180^\circ}{\pi} \arcsin(\varepsilon)(1 + 0.1547\varepsilon^2) \tag{5.18}$$

US EPA (2000) [9] has also mentioned the Mardia method [12], which can be given as

$$\sigma_\theta = \frac{180^\circ}{\pi} \sqrt{-\ln(1 - \varepsilon^2)} \tag{5.19}$$

A similar trending effect is also applicable for wind direction standard deviations, but it is not as important an issue as the turbulence intensity for wind resource assessment. Therefore the issue is not discussed here.

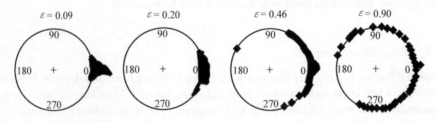

Figure 5.5 Random wind direction measurements. Same mean direction (zero degree) with different spreads of the data points

5.3   Mean Wind Speed of the Whole Time Series

The mean wind speed of the whole time series, V_{ave}, can be calculated directly from the mean wind speed time series as given in the following:

$$V_{ave} = \frac{1}{n}\sum_{j=1}^{n} v_j \tag{5.20}$$

where v_j is the jth mean wind speed record in the data series usually covering a whole year or a number of whole years and n represents the number of data records.

The standard deviation:

$$\sigma_V = \sqrt{\frac{\sum_{j=1}^{n}(v_j - V_{ave})}{n}} \tag{5.21}$$

It should be emphasised that Equation (5.20) cannot be used directly to calculate the wind power density (which is proportional to the cubic wind speed) from Equation (4.4), as introduced in Section 4.1. The average wind speed representing the power content of the wind should follow:

$$V'_{ave} = \left(\frac{1}{n}\sum_{j=1}^{n} v_j^3\right)^{1/3} \tag{5.22}$$

Take three mean wind speed records of 5 m/s, 7 m/s and 9 m/s as an example, as shown in Table 5.1. The mean wind speed following Equation (5.20) is 7.0 m/s, giving a mean wind power density of 210.1 W/m^2, whereas the mean of the three wind power densities (76.6 W/m^2, 201.1 W/m^2 and 446.5 W/m^2) is actually 244.4 W/m^2. Therefore, the power content of the wind speed should follow Equation (5.22).

Normally, mean wind speed time series are binned into n equal-sized wind speed ranges (for instance, 0~1 m/s, 1~2 m/s, 2~3 m/s, etc.). Taking f_i as the frequency of occurrence in bin i, the mean wind speed can be approximated as

$$V_{ave} = \frac{\sum_{i=1}^{n} f_i V_i^M}{\sum_{i=1}^{n} f_i} \tag{5.23}$$

where V_i^M represents the median value of the wind speed bin i.

Table 5.1   Example of mean wind speed calculations

j	1	2	3	Mean (5.20)	Power mean (5.22)
WS (m/s)	5.0	7.0	9.0	7.0	7.4
WPD (W/m^2)	76.6	210.1	446.5	210.1 (wrong)	244.4

5.4   Weibull Distribution

Table 5.1 has exhibited the importance of wind speed distribution to wind power availability. Mean wind speed distributions over a long time usually have a long right tail, that is a
positive skewness. This means that wind speed values are bounded to one side. Also, wind
speed values cannot be negative. Given the features of mean wind speed distributions, the
two-parameter Weibull distribution, which is named after the Swedish mathematician Waloddi
Weibull (1887–1979), is found to be the most suitable description.

5.4.1   Weibull Probability Density Function

The Weibull distribution gives a graphical representation of how often the wind blows at a
certain speed over a long period of time. The two-parameter probability density function of
the Weibull distribution is given as

$$f(v) = \frac{k}{a}\left(\frac{v}{a}\right)^{k-1} \exp\left[-\left(\frac{v}{a}\right)^{k}\right] \tag{5.24}$$

where the constant k is the shape/form parameter and constant a is the scale parameter.

The dimensionless shape parameter k has a marked effect on the behaviour of the distribution,
as shown in Figure 5.6. When $k = 1.0$, the Weibull function reduces to the exponential function;
when k is about 3.5, it closely approximates the normal distribution; between 1.0 and 3.5, a
larger k value will give a narrower function curve, that is the occurrences of the wind speed are
more concentrated around the mean; when $k = 2.0$, it is also called the Rayleigh distribution.
The constant a is given in m/s and principally is proportional to the mean wind speed of the
whole time series.

The peak of the Weibull function is on the left-hand side of the mean, which means that the
wind speeds are more likely to be lower than the mean. The long tail on the right approaches
zero infinitely as the wind speed increases.

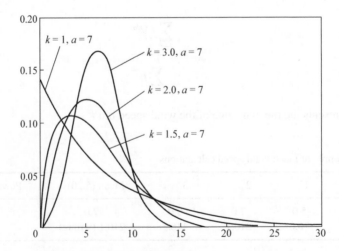

Figure 5.6   Two-parameter Weibull probability density function

According to the definition of Equation (5.5), the annual mean wind speed can be given as

$$V_{ave} = \int_0^{+\infty} vf(v)dv \tag{5.25}$$

Let

$$x = \left(\frac{v}{a}\right)^k \tag{5.26}$$

Then

$$dv = \frac{a}{k}x^{(1/k-1)} \tag{5.27}$$

Substituting Equation (5.25) with Equations (5.24) and (5.27), we can get

$$V_{ave} = a \int_0^{+\infty} e^{-x}x^{1/k}dx \tag{5.28}$$

Equation (5.28) can be rewritten in the form of a gamma function[1] as

$$V_{ave} = a\Gamma\left(1 + \frac{1}{k}\right) \tag{5.29}$$

Therefore, the Weibull distribution is determined as long as two of the three parameters, V_{ave}, a and k, are known.

Higher central moments of the Weibull distribution V_n are given by

$$V_n = a^n\Gamma\left(1 + \frac{n}{k}\right) \tag{5.30}$$

Applying Equation (5.3), the standard deviation of the Weibull distribution can be derived as follows:

$$\sigma_V = \sqrt{V_2 - V_{ave}^2} = a\sqrt{\Gamma\left(1 + \frac{2}{k}\right) - \Gamma^2\left(1 + \frac{1}{k}\right)} \tag{5.31}$$

The Weibull distribution is usually fitted with wind data measurements whose occurrences are sorted into sector-wise wind speed bins, as shown in Figure 5.7. The fitted Weibull distributions for each direction sector are then used to statistically model the wind resource rather than directly using wind data measurements. How well the Weibull distribution fits wind data measurements must be checked. Wind data screening and filtering may be necessary in order to improve the fitting quality (see Section 5.5).

5.4.2 Weibull Cumulative Distribution Function

The cumulative distribution function can be used to calculate the probability of the wind speed occurring in a certain wind speed range. It is given by

$$F(v_x) = P(v < v_x) = \int_0^{v_x} f(v)dv = 1 - \exp\left[-\left(\frac{v_x}{a}\right)^k\right] \tag{5.32}$$

---

[1] The gamma function, $\Gamma(x) = \int_0^{+\infty} e^{-x}x^{t-1}dt$, has many useful algorithms, such as $\Gamma(t+1) = t\Gamma(t)$, $\Gamma\left(\frac{1}{2}\right) = \pi$ and $\Gamma(1) = 1$. The gamma function in MS Excel is = EXP(GAMMALN(x)).

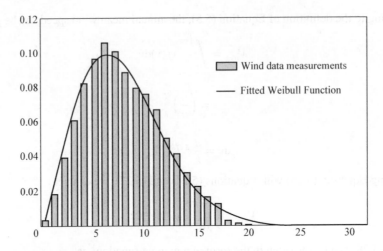

Figure 5.7   Weibull PDF fitted with wind data measurements for all direction sectors

Thus, the probability of the wind speed in a certain wind speed range[2] is

$$P(v_1 < v < v_2) = F(v_2) - F(v_1) \tag{5.33}$$

$$P(v_1 < v < v_2) = \exp\left[-\left(\frac{v_1}{a}\right)^k\right] - \exp\left[-\left(\frac{v_2}{a}\right)^k\right] \tag{5.34}$$

The frequency of the wind speed higher than a certain wind speed is

$$P(v > v_x) = 1 - F(v_x) = \exp\left[-\left(\frac{v_x}{a}\right)^k\right] \tag{5.35}$$

The Weibull CDF fitted with wind data measurements is shown in Figure 5.8.

___

Example 5.1

The cut-in and cut-out wind speeds of a wind turbine are 4 m/s and 25 m/s and the Weibull parameters of the wind at hub height are $k = 2.2$ and $a = 9$ m/s. How many hours in a year would the wind turbine be producing electricity? What is the probability of the wind speed exceeding 30 m/s?

Solution

According to Equation (5.34), the probability of the wind speed occurring in the power-generating wind speed range (4 m/s to 25 m/s) can be calculated by

$$P(4 < v < 25) = \exp\left[-\left(\frac{4}{9}\right)^{2.2}\right] - \exp\left[-\left(\frac{25}{9}\right)^{2.2}\right] \approx 0.8453$$

___

[2] In MS Excel, the function is given as: =WEIBULL(v_2,k,a,TRUE)-WEIBULL(v_1,k,a,TRUE).

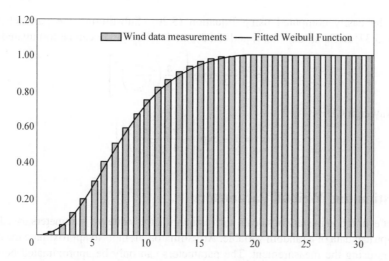

Figure 5.8   Weibull CDF fitted with wind data measurements (same data as in Figure 5.7) of all direction sectors

Multiplying the result above with the number of hours in a year, we get the annual available hours for the wind turbine to generate electricity as

$$0.8453 \times 8760 \text{ h} \approx 7405 \text{ h}$$

The probability of the wind speed exceeding 30 m/s in a year can be computed from Equation (5.35):

$$P(v > 30) = \exp\left[-\left(\frac{30}{9}\right)^{2.2}\right] \approx 0.00000073$$

It is clear that the wind speed exceeding 30 m/s is really rare. This formula, however, cannot be used to estimate the 50-year extreme wind speed, which in theory follows a different probability distribution and should be studied separately (see Section 5.6).

5.4.3   Rayleigh Distribution

The Rayleigh distribution is the Weibull distribution when the shape parameter k = 2, representing the average case of wind distributions around the world. The Rayleigh distribution is assumed in the IEC standards [4] to define wind turbine load cases.

Taking $k = 2$ in Equation (5.29) gives

$$V_{ave} = a\Gamma\left(\frac{3}{2}\right) \tag{5.36}$$

Applying the algorithms of the gamma function, the a parameter can be deduced as

$$a = \frac{2V_{ave}}{\sqrt{\pi}} \tag{5.37}$$

V_{ave} can be easily computed using Equation (5.20). Substituting the parameter a with Equation (5.37) in Equations (5.24) and (5.32), the Rayleigh PDF can be formulated as

$$f(v) = \frac{\pi}{2} \frac{v}{V_{ave}^2} \exp\left[-\frac{\pi}{4}\left(\frac{v}{V_{ave}}\right)^2\right] \tag{5.38}$$

and the Rayleigh CDF is

$$F(v) = 1 - \exp\left[-\frac{\pi}{4}\left(\frac{v}{V_{ave}}\right)^2\right] \tag{5.39}$$

5.5 Estimating Weibull Parameters

Parameter estimation in statistics deals with estimating the values of parameters based on measured/empirical data of a random variable. An estimator attempts to approximate the unknown parameters using the measurement. The parameters can only be approximated because the measured data as a sampling space cannot cover the entire population, that is the wind data measurement can never cover the wind entirely. Therefore, the quality of the parameter estimates has to be checked and evaluated mathematically (see Section 5.5.6).

Many methods have been developed to estimate the Weibull parameters. Five of these methods (the linear regression method [14], the mean-standard deviation method, the maximum likelihood method [15], the median method [16] and the power density method [17]) that are most commonly used in wind analysis are introduced in this chapter.

5.5.1 Linear Regression Method

First, the wind speed measurements are sorted into wind speed bins and counted in order to calculate the probability of occurrence in each bin. Rewrite Equation (5.32) as

$$1 - F(v_i) = \exp\left[-\left(\frac{v_i}{a}\right)^k\right] \tag{5.40}$$

where v_i denotes the median of the wind speed bin i.

Applying the logarithmic operation twice on both sides of the equation, we get:

$$\ln\{\ln[1 - F(v_i)]\} = k\ln(v_i) - k\ln(a) \tag{5.41}$$

Let

$$y = \ln\{\ln[1 - F(v_i)]\} \tag{5.42}$$

$$x = \ln(v_i) \tag{5.43}$$

$$c = k\ln(a) \tag{5.44}$$

Thus a linearised relation can be achieved:

$$y = kx - c \tag{5.45}$$

If the wind speed follows a Weibull distribution, it should present a linear behaviour described by Equation (5.45) and the slope of the line is equal to the Weibull shape parameter (this is why the shape parameter is often referred to as the slope parameter). The least square linear regression method is usually used graphically to fit the line and derive the Weibull parameters.

5.5.2 Mean-Standard Deviation Method

The mean and standard deviation of wind data can be calculated from Equations (5.20) and (5.21). Thus one can derive the Weibull k parameter by combining Equations (5.29) and (5.31), as expressed by

$$\left(\frac{\sigma_V}{V_{ave}}\right)^2 = \frac{\Gamma\left(1 + \frac{2}{k}\right)}{\Gamma^2\left(1 + \frac{1}{k}\right)} - 1 \tag{5.46}$$

It is usually acceptable to calculate k approximately by

$$k = \left(\frac{\sigma_V}{V_{ave}}\right)^{-1.086} \tag{5.47}$$

Once k is determined, the a parameter can be calculated from Equation (5.29).

5.5.3 Maximum Likelihood Estimate Method

In statistics, the maximum likelihood estimate (MLE) is one of the most commonly used techniques for estimating the parameters of a statistic model. MLE is favourable because of the large sample properties like consistency and asymptotic unbiasedness and the smallest statistical error. Consistency means that as the number of observations goes to infinity the MLE estimator converges in probability to its true value, approximately unbiased. Asymptotic unbiasedness and the smallest statistical error in general means that the MLE estimator is approximately optimal.

For the Weibull distribution, the MLE estimator is obtained by maximising the objective function as

$$L(v_1, v_2, \ldots, v_n | a, k) = \prod_{j=1}^{n} \left(\frac{k}{a}\right)\left(\frac{v_j}{a}\right)^{k-1} \exp\left[-\left(\frac{v_j}{a}\right)^k\right] \tag{5.48}$$

which results in the log-MLE estimator:

$$\ln L = n \ln \left(\frac{k}{a}\right) + (k-1) \sum_{j=1}^{n} \ln \left(\frac{v_j}{a}\right) - \sum_{j=1}^{n} \left(\frac{v_j}{a}\right)^k \tag{5.49}$$

Differentiating Equation (5.49) with respect to k and a in turn and equating them to zero to obtain the following estimating equations:

$$\frac{\partial \ln L}{\partial k} = \frac{n}{k} + \sum_{j=1}^{n} \ln v_j - \frac{1}{a} \sum_{j=1}^{n} v_j^k \ln v_j = 0 \tag{5.50}$$

$$\frac{\partial \ln L}{\partial a} = -\frac{n}{k} + \frac{1}{a^2} \sum_{j=1}^{n} v_j^k = 0 \tag{5.51}$$

gives

$$a = \left(\frac{1}{n} \sum_{j=1}^{n} v_j^k \right)^{1/k} \tag{5.52}$$

$$\frac{\sum\limits_{i=1}^{n} v_j^k \ln v_j}{\sum\limits_{i=1}^{n} v_j^k} - \frac{1}{k} - \frac{1}{n} \sum_{i=1}^{n} \ln v_j = 0 \tag{5.53}$$

These equations may be solved by the use of iterative procedures.

The MLE method is usually recommended due to its statistical properties, but it is very sensitive (similar to other statistical methods) to outliers, which are generally the result of mistakes in measuring and recording data, and becomes unreliable [18].

5.5.4 Medians Method

In order to reject or reduce the effect of outliers and provide a better fit to the majority of data, robust estimators may be good alternatives [19]. A robust and rather efficient alternative for the MLE estimator is the method of medians proposed by He and Fung (1999) [16]. Kantar et al. (2013) have examined [18] the medians method for wind energy applications and agrees with its robustness.

From logarithms of Equation (5.24), we can derive

$$\ln f(v) = \ln k + \ln a + \frac{(k-1)}{a} \ln v - \left(\frac{v}{a} \right)^k \tag{5.54}$$

The score function of the Weibull distribution is then given by

$$\varphi(v) = \frac{\partial \ln f(x)}{\partial \beta(a,k)} \tag{5.55}$$

The score function is calculated as

$$\varphi(v) = \begin{cases} ka \left(1 - \left(\frac{v}{a} \right)^k \right) \\ \frac{1}{k} \left(1 + \left(1 - \left(\frac{v}{a} \right)^k \right) \ln \left(\frac{v}{a} \right)^k \right) \end{cases} \tag{5.56}$$

The medians method equates the sample median of $\varphi(v_j)$ with the population median. Because $y = \left(\frac{x}{a} \right)^k$ has an exponential distribution with mean one, the following equations can be achieved:

$$\text{Median}_i \left\{ \left(\frac{v_i}{a} \right)^k \right\} = \ln 2 \tag{5.57}$$

$$\text{Median}_i \left\{ \left(1 - \left(\frac{v_i}{a} \right)^k \right) \ln \left(\frac{v_i}{a} \right)^k \right\} \approx -0.51 \tag{5.58}$$

From Equation (5.57), it follows that

$$a = \frac{\text{Median}_i\{v_i\}}{(\ln 2)^{1/k}} \qquad (5.59)$$

Taking Equation (5.59) into Equation (5.58), then (5.58) can be solved iteratively for k.

5.5.5 Power Density Method

Akdag and Ali (2009) [17] suggest a new method to estimate the Weibull parameters based on the unique applications of wind energy, that is the energy content (to the third power of the wind speed) of the Weibull distribution must be the same as the actual data distribution.

According to Akdag and Ali (2009) [17], an energy pattern factor, E_{PF}, is defined as

$$E_{PF} = \frac{E[v^3]}{E^3[v]} = \frac{\dfrac{1}{n}\displaystyle\sum_j^n v_j^3}{\left(\dfrac{1}{n}\displaystyle\sum_j^n v_j\right)^3} = \frac{\Gamma\left(1+\dfrac{3}{k}\right)}{\Gamma^3\left(1+\dfrac{1}{k}\right)} \qquad (5.60)$$

The k parameter can be estimated by numerically solving the last two equating items in Equation (5.60), or by using the approximate formula:

$$k = 1 + \frac{3.69}{E_{PF}^2} \qquad (5.61)$$

Once k is determined, the a parameter can be calculated from Equation (5.29).

Taking a similar idea to the equal energy content, WAsP implements an engineering approach to estimate the Weibull parameters. The assumption of an equal energy content between wind data and the Weibull distribution gives rise to

$$a^3 \Gamma\left(1+\frac{3}{k}\right) = \frac{1}{n}\sum_j^n v_j^3 \qquad (5.62)$$

Thus,

$$a = \left[\frac{\dfrac{1}{n}\displaystyle\sum_j^n v_j^3}{\Gamma\left(1+\dfrac{3}{k}\right)}\right]^{1/3} \qquad (5.63)$$

Solving Equation (5.63) requires another equation, which is constructed on the assumption that the frequencies of the speeds higher than the mean must be the same in both (wind data and Weibull) distributions. This assumption makes sense for wind energy applications because a wind turbine produces the majority of the energy at wind speeds higher than the mean and low wind speeds are usually statistically less reliable. From Equation (5.35) we know that the probability of the wind speed exceeding the mean is

$$P_{ave}(v > V_{ave}) = \exp\left[-\left(\frac{V_{ave}}{a}\right)^k\right] \qquad (5.64)$$

Substituting V_{ave} with Equation (5.29) and transforming the equation, we get

$$k = \frac{\ln(-\ln(P_{ave}))}{\ln\left(\Gamma\left(\frac{1}{k}+1\right)\right)} \tag{5.65}$$

The probability of occurrence, P_{ave}, can be derived from wind data. Equation (5.65) is then solved iteratively to achieve k.

5.5.6 Quality of the Weibull Fit

In statistics, the coefficient of determination, R^2, indicates how well data points fit a statistical model as the proportion of total variation of outcomes explained by the model. It is the ratio of the explained variation to the total variation, representing the percent of the data that is closest to the line of best fit.

R^2 takes on values between 0 and 1; the higher the R^2 the more useful is the model. An R^2 of 1 indicates that the statistical model perfectly fits the data.

From its definition, we can expect R^2 to be a good candidate for checking how well the Weibull distribution fits wind data measurements. Different parameter estimate methods mentioned previously may lead to slightly different R^2 values.

The general form of coefficient of determination, R^2, can be given by

$$R^2 = 1 - \frac{SS_{err}}{SS_{tot}} \tag{5.66}$$

$$SS_{err} = \sum (X_i - x_i)^2 \tag{5.67}$$

$$SS_{tot} = \sum (X_i - \overline{X})^2 \tag{5.68}$$

where X_i is the frequency of observed wind data in the wind speed bin i, x_i the frequency estimated with fitted Weibull distribution and $\overline{X}$ the mean of X_i values. Therefore, SS_{err} denotes the sum of squares of the error/residual, measuring the deviations of observations from their predicted values, while SS_{tot} is the total sum of squares.

The coefficient of determination over the wind speed range of a power curve is of greater importance because wind speeds below cut-in and above cut-out are not considered when it comes to calculating power production of the wind turbine. In general, R^2 of the wind speed Weibull fit should be no less than 0.99 to be considered accurate. Some wind climates may not fit the Weibull distribution very well and may even show double peaks, leading to biased production estimates. The misfit may be caused by the portions of wind data outside the power curve range being too large. Excluding them in the Weibull fitting procedure may improve the fits dramatically in the more critical power-producing wind speed range. When the Weibull fit is considered unsatisfactory, it is still possible to evaluate the bias on production estimates by calculating the difference between the production estimates based on the data and the Weibull distributions.

The root mean square error (RMSE) (often referred to as the root mean square deviation (RMSD)) is also often employed to compare the fitness of different Weibull fitting methods. It represents the sample standard deviation of the differences between predicted values and observed values, aggregating the magnitudes of the errors in the statistic model for various times into a single measure of predictive power. RMSD is a good measure of accuracy, but only to compare errors of different models for a particular variable and not between variables, as it is scale-dependent [20]. RMSD also takes on values between 0 and 1, but smaller RMSD generally means more accurate modelling.

RMSD can be given as

$$RMSE = \sqrt{\frac{1}{n} \sum (X_i - x_i)^2}$$ (5.69)

Example 5.2

A wind data series for a whole year is summarised in the three columns from the left in Table 5.2, only taking into consideration the wind speed range that overlaps that of the power curve (4–25 m/s). The frequency calculated from the Weibull fit is given in the last column. The wind speed bins represent a wind speed range of $i \pm 0.5$ m/s. How well does the Weibull distribution fit the wind data?

Solution

We can use the coefficient of determination, R^2, to evaluate the goodness of the Weibull fit. From Equations (5.67) and (5.68), it can be determined that the total sum of squares,

Table 5.2 Frequency of mean wind speed data series

Wind speed bin i (m/s)	Counts	Data frequency X_i	Weibull frequency x_i
4	10776	0.08254	0.08550
5	15829	0.09630	0.09506
6	21364	0.10549	0.09876
7	26631	0.10038	0.09710
8	31260	0.08822	0.09103
9	35402	0.07894	0.08178
10	39378	0.07578	0.07064
11	42867	0.06650	0.05880
12	45517	0.05051	0.04726
13	47708	0.04176	0.03671
14	49341	0.03112	0.02760
15	50543	0.02291	0.02009
16	51416	0.01664	0.01418
17	52065	0.01237	0.00970
18	52237	0.00328	0.00644
19	52297	0.00114	0.00414
20	52331	0.00065	0.00259
21	52357	0.00050	0.00157
22	52392	0.00067	0.00093
23	52412	0.00038	0.00053
24	52431		

$SS_{tot} = 0.0002617$ and the sum of squares of error, $SS_{err} = 0.0343724$. Thus, R^2 can be calculated from (5.66), $R^2 = 0.992$. As a result, it can be concluded that the Weibull fits very well with the wind data.

5.6   Extreme Wind Statistics

Fifty-year extreme wind and gusts play an important role in the load case design of wind turbines. IEC 61400-1 (2005) [4] defines reference wind conditions (see Section 4.4.4, Wind Turbine Selection) on the basis of the 50-year 10-minutes extreme wind speed. Statistically estimating extreme winds is therefore an important topic in wind resource assessment.

The probability of extreme and rare wind events follows extreme value distributions, which are different from those of mean wind speeds. The study of extreme and rare events focuses on the asymptotic tail of the distribution of mean values. This section will introduce three important statistical methods for extreme winds: the Gumbel method, the peak-over-threshold model and the independent storm method.

5.6.1   Independent Extreme Wind Events

Generally speaking, the study of extreme winds is the parameter estimate of extreme value distributions based on a set of independent largest order-ranked wind speed samples, which can be given by

$$\{v_1, v_2, v_3, v_j, \ldots, v_n\} \quad \text{and} \quad (v_1 < v_2 < v_3 < v_j < \cdots < v_n) \tag{5.70}$$

Extreme wind speeds may arise from a variety of physically distinct meteorological mechanisms, including tropical cyclones, thunderstorms and depressions. Each of these independent mechanisms produces a series of isolated and independent extreme events that can be described by a Poisson process, a stochastic process that counts the number of events and the time that these events occur in a given time interval. Multiple independent physical mechanisms (mixed climate) will give rise to overlapping of the Poisson processes. However, in wind engineering considering extreme wind statistics, a simple wind climate with only one possible dominating mechanism, where the contributions from other mechanisms are negligible, is often assumed. The assumption is usually reasonable for sites in temperate latitudes, with depressions being the dominant mechanism for the UK [21] and typhoons for the western North Pacific [22]. If no single dominant mechanism can be identified, separate statistical analysis of extreme winds caused by each important mechanism should be applied.

The extreme wind speed, V_T, can be defined as a wind speed that on average is exceeded once during the recurrence period, T [23]. A Poisson process is implicitly assumed in the definition. For a Poisson process, the time between each pair of consecutive events has an exponential distribution and each of these interarrival times is assumed to be independent of other interarrival times.

In estimating extreme wind speeds for 50 years, we usually also assume that wind speeds for different years have identical and independent statistical properties, even though the validity of this assumption is debatable [24], that is extreme wind events occur at a constant rate, so wind observations of only a few years can be extrapolated to 50 years. Such a process is called a homogeneous Poisson process, which can be characterised by a rate parameter λ, also known as the intensity.

5.6.2   Gumbel Method

In statistics, the focus on average behaviour of a stochastic process is called the central limit theorem, while the focus on extreme and rare events is called extreme value theory, also known as the generalised extreme value (GEV) distribution or the Fisher–Tippett theorem. The Gumbel distribution [25] belongs to the family of GEV distributions (see Table 5.3) [26], which can be mathematically defined as follows [27]:

$$F(x) = \exp\left[-\left(1 - k\left(\frac{x - \beta}{a}\right)\right)^{1/k}\right], \quad k \neq 0 \tag{5.71a}$$

$$F(x) = \exp\left[-\exp\left(1 - \left(\frac{x - \beta}{a}\right)\right)\right], \quad k = 0 \tag{5.71b}$$

where β $(-\infty < \beta < \infty)$, a $(0 < a < \infty)$ and k $(-\infty < k < \infty)$ are the location, scale and shape parameters respectively. The value of the shape factor k defines the GEV distribution to be of Type 1 (Gumbel), Type 2 (Frechet) or Type 3 (Weibull), as shown in Table 5.3.

The use of the Gumbel distribution is especially well-established for modelling extremes of natural phenomena, including floods and wind speeds, and becomes the engineering favourite due to its linear property.

From Equation (5.71a), the cumulative probability function of extreme wind speeds for one year can be given by the Gumbel distribution ($k = 0$):

$$F(V) = \exp\left[-\exp\left(1 - \left(\frac{V - \beta}{a}\right)\right)\right] \tag{5.72}$$

Due to the assumed independence between different years, the CDF for a T year period is

$$F_T(V) = \exp\left[-T\exp\left(1 - \left(\frac{V - \beta}{a}\right)\right)\right] = \exp\left[-\exp\left(-\frac{V - \beta - \ln T}{\alpha}\right)\right] \tag{5.73}$$

Thus

$$-\ln(-\ln(F_T)) = \frac{V_T - \beta - \ln T}{\alpha} \tag{5.74}$$

Let

$$y = -\ln(-\ln(F_T)) \tag{5.75}$$

Table 5.3   GEV types and their properties [27]

Name	Type	k	Curvature	Lower bound	Upper bound
Gumbel distribution	1	$k = 0$	Linear	None	None
Frechet family	2	$k < 0$	Negative	$a/k + u$	None
Weibull family	3	$k > 0$	Positive	None	$a/k + u$

Then

$$V_T = \alpha y + \frac{\beta}{\alpha} + \alpha \ln T \qquad (5.76)$$

We can see from Equation (5.76) that V_T and y can be plotted in a straight line. In practice, the first step is to identify independent maxima of available wind speed time series (e.g. annual extreme values). These maxima are then sorted (order-ranked) in ascending order according to Equation (5.70). In the standard Gumbel method, the cumulative probability, F_j, is assigned to each ranked v_j given by the plotting position [28]:

$$F_j = \frac{j}{1+n} \qquad (5.77)$$

Let

$$y_j = -\ln(-\ln(F_j)) \qquad (5.78)$$

Secondly, the sorted wind speed maxima are plotted against y_j and data that follow a Gumbel distribution will form in a straight line, as shown in Figure 5.9. Lastly, by extrapolating such a graph, extreme wind speeds for a given return period (usually 50 years) can be easily extrapolated. Now we simply need to identify the position on the extrapolated graph for the desired extreme value. There are two methods used to identify the location.

In the first method, the cumulative probability of not exceeding the extreme value for 50-year period, F_T, is directly calculated to find the corresponding value of y in Equation (5.75). The 50-year extreme wind speed is found where the extrapolated straight line

$$V_T = ay + b \qquad (5.79)$$

crosses the value of F_T. F_T depends on the number of extreme wind speeds per year, m, taken into account:

$$F_T = 1 - \frac{1}{m} \qquad (5.80)$$

For instance, if annual wind speed maxima (to ensure the independence) are used, then the number of extreme wind speeds for 50 years, $m = 50$, gives $F_T = 1 - 1/50 = 0.98$. However,

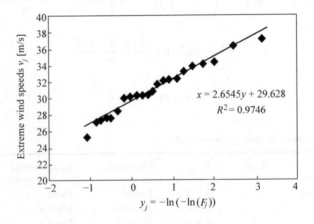

Figure 5.9   Gumbel regression of extreme wind speeds

if a time series is much shorter than the required return period, which is usually the case in wind energy, the series of annual extreme values will be too small for a meaningful analysis. Therefore, multiple extreme values per year are more desirable in practice.

In the second method, T denotes the period of available wind speed time series given as the number of whole years and V_T is the maximum value on the regression straight line, given by

$$V_T = \alpha \max\{y_1, y_2, \ldots, y_j, \ldots, y_n\} + \frac{\beta}{\alpha} \tag{5.81}$$

Therefore, the 50-year extreme wind, V_{50}, can be given as

$$V_{50} = V_T + \alpha \ln \frac{50}{T} \tag{5.82}$$

Example 5.3

The top 23 extreme wind speed values are taken from a 2-year wind speed time series and ranked in ascending order, as given by Table 5.4. F_j and y_j have also been calculated according to Equations (5.77) and (5.78). What is the estimated 50-year extreme wind speed?

Table 5.4 Extreme wind speeds from a 2-year data series prepared for Gumbel regression

Observed maximum wind speeds v_j (m/s)	Rank j	CDF F_j	$y_j = -\ln(-\ln(F_j))$
25.5	1	0.04	−1.2
27.2	2	0.08	−0.9
27.5	3	0.13	−0.7
27.8	4	0.17	−0.6
27.9	5	0.21	−0.5
28.5	6	0.25	−0.3
29.9	7	0.29	−0.2
30.2	8	0.33	−0.1
30.2	9	0.38	0.0
30.3	10	0.42	0.1
30.3	11	0.46	0.2
30.5	12	0.50	0.4
30.8	13	0.54	0.5
31.7	14	0.58	0.6
32.1	15	0.63	0.8
32.3	16	0.67	0.9
32.4	17	0.71	1.1
33.2	18	0.75	1.2
33.9	19	0.79	1.5
34.2	20	0.83	1.7
34.4	21	0.88	2.0
36.3	22	0.92	2.4
37.2	23	0.96	3.2

Solution

Here y_i is plotted against v_i and the least-square linear regression is performed to fabricate Figure 5.9.

Comparing the regression line in Figure 5.9 with Equation (5.76), we find that with $T = 2$

$$\alpha = 2.6545$$

and

$$\frac{\beta}{\alpha} + \alpha \ln 2 = 29.628$$

Thus, $\beta = 73.764$. From Equation (5.81), the 2-year extreme wind speed can be computed and $V_2 = 38$ m/s. The 50-year extreme wind speed is then calculated from Equation (5.82) as the following:

$$V_{50} = V_2 + 2.2645 \ln \left(\frac{50}{2} \right) = 46.6 \ \text{m/s}$$

The extrapolated graph of the Gumbel distribution is given in Figure 5.10. Using the first method, the position of y for the 50-year extreme wind speed can be calculated directly from Equations (5.80) and (5.75) as

$$y = -\ln(-\ln(F_T)) = -\ln \left(-\ln \left(1 - \frac{1}{50 \times 23/2} \right) \right) = 6.38$$

Taking $y = 6.38$ into the regression line in Figure 5.10, we can get the same result.

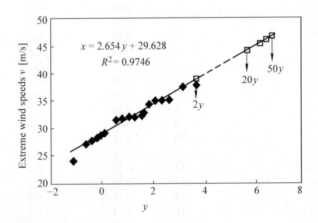

Figure 5.10 Extrapolation of the Gumbel regression in Figure 5.9

The example above demonstrates the procedure of the classical Gumbel method, but there are a few noteworthy theoretical arguments that should be made known. A profound theoretical discussion can be found in Harris (1996) [21].

Because the standard Gumbel method is graphically based, it is easy to find whether or not the plot of y versus extreme wind speeds gives a straight line. In practice, the curvature may arise from the failure to fit the GEV type I asymptote where type II and especially type III may be more appropriate, or because the data points come from a mixed climate and Poisson process [21].

Statistical independence of the extreme wind speeds is presumed for the Gumbel distribution. Wind speeds from the same storm are very unlikely to be statistically independent events. For instance, typhoons usually give two wind speed peaks, one for each of the two passages of the eyewall. This problem is usually tackled by placing a 'dead time' window between two neighbouring wind speeds on the time series to avoid multiple counting of consecutive exceedances from single storms. The window may vary from a few hours to a few days depending on the time it takes for the dominant extreme wind system to pass [22].

To reduce the effect of less important extreme wind mechanisms, improve the statistical independence and make a meaningful statistical analysis, the number of extreme wind speeds used for the regression should usually range from 10 to 50, achieved by adjusting the 'dead time' window. This approach is adopted by WindPRO.

Harris (1996) [21] also argues that the classical least-squares technique used in the standard Gumbel method for regression is not really suitable for Gumbel plotting of extreme wind speeds, because the classical least-squares method presumes standard deviations associated with the order statistics to be of a similar magnitude. For extreme value data, however, the standard deviation associated with the plotted probability ordinates varies systematically, being largest for the largest value. Consequently, the classical least-squares method places too much weight on the deviations of the largest extreme values.

Graphically, the Gumbel plot should try to better fit the largest 3 to 5 extreme values. Mathematically, Harris (1996) [21] proposes a weighted least-squares method by revising the plotting positions. The weight is inversely proportional to the corresponding rank's standard deviation. For convenience, the ordering procedure is the reverse, $j = 1$ for the largest sample down to $j = n$ for the smallest. Instead of Equation (5.77), the probability density of the jth ranked value is given by

$$P_j(z) = \frac{n!}{(j-1)!(n-j)!} z^{n-j} (1-z)^{j-1} \tag{5.83}$$

Thus the mean plotting position for the jth ranked value is given by

$$\bar{y}_j = \frac{n!}{(j-1)!(n-j)!} \int_0^1 -\ln(-\ln(z)) z^{n-j} (1-z)^{j-1} dz \tag{5.84}$$

where z denotes the Gumbel cumulative probability distribution, $F(V)$, as given in Equation (5.72).

The parameters of a straight line are obtained by minimising the quantity S^2 as

$$S^2 = \sum_{j=1}^n w_j (\bar{y}_j - aq_j - b)^2 \tag{5.85}$$

which is subjected to

$$\sum_{j=1}^n w_j = 1 \quad \text{and} \quad w_j = \frac{1/\sigma_j^2}{\displaystyle\sum_{j=1}^n 1/\sigma_j^2} \tag{5.86}$$

where q_j is the value of the squared wind speed, which gives greater weight to the higher wind speeds, a and b are the parameters of the straight line and w_j and σ_v are the weight and standard deviation for the jth ranked sample respectively.

5.6.3 Peaks-Over-Threshold Method

In contrast to the Gumbel method, which only utilises a few extreme wind speed samples, the peak-over-threshold (POT) method takes into account all measurements above a chosen threshold, that is peaks over threshold. This approach increases the number of samples included in the analysis and thereby reduces the statistical uncertainty of quantile variances introduced in the Gumbel method [29]. It is especially more attractive when the wind data series in question is much shorter than the return period.

The POT method has been described in Pickands (1975) [30] and the generalised Pareto distribution (GPD) has been found to be the appropriate distribution method for exceedances over a threshold under the same basic conditions as those for the Gumbel distribution. Provided also that the up-crossings of the threshold are Poisson and thus the exceedances themselves are independent, the GPD and GEV are closely related and even share the same shape factor, k.

The GPD has a cumulative distribution function defined by [29]

$$F(x) = 1 - \left[1 - \left(1 - \frac{k}{a}(1 - \xi)\right)\right]^{1/k}, \quad k \neq 0 \qquad (5.87a)$$

$$F(x) = 1 - \exp\left[-\frac{x - \xi}{\alpha}\right], \quad k = 0 \qquad (5.87b)$$

where ξ, α and k denote the location (the given threshold), scale and shape parameters respectively; ξ is subjected to $\xi \leq x \leq \xi + \alpha/k$ for $k > 0$ and $\xi \leq x < \infty$ for $k \leq 0$.

Defining the crossing rate λ as the expected unity number of peaks above the threshold per year, the extreme value in T years can be solved by

$$x_T = \xi + \frac{a}{k}[1 - (\lambda T)^{-k}] \qquad (5.88)$$

The problem of estimating the extreme value then devolves to estimating the GPD parameters α and k for a given threshold ξ [29], which can be achieved by classical methodologies in statistics such as the maximum likelihood.

The POT method inevitably raises the issue of event independence and the optimal choice of threshold level. To make sure that the peaks used to determine the Pareto parameters are statistically independent, one usually has to make a compromise between long 'dead time' and a large enough sample from a limited number of time series. After fitting the Pareto distribution, quality checks should be carried out to ensure the statistical independence and quality of the fit (the same applies to the Gumbel method). Brabson (2000) [29] presents good reviews on the event independence, threshold and identification of mixed climates for using the POT method.

A similar approach to POT is the method of independent storms (MIS) method, which defines the beginning of the 'lulls' in the record by the downward crossings of the wind speed below a chosen threshold. It uses the actual values of storm maxima and ignores other smaller peaks within a storm and the wind speed records between each pair of lulls are considered as a part of an independent storm [31]. The MIS method, as a data selection technique to ensure event independence, can be applied to both the GEV and the GPD methods.

5.6.4   Extreme Wind Gusts

Wind gusts are characterised by a rapid increase in wind speed and a subsequent decrease in a short period of time, usually 3 seconds as defined in IEC 61400-1 [4]. They can also be defined as the short-term turbulent fluctuations in the wind speed for duration of a few seconds at most [32].

Fast-sampled wind data are more likely to be unavailable for wind engineers to derive extreme gusts directly, and thus a statistical approach based on the time-averaged wind data series has to be applied in most cases. In an averaging interval (usually 10 minutes or 1 hour), the wind speed is commonly assumed to be following a normal distribution, a good assumption especially for higher wind speeds. Thus the gust speed, V_{gust}, can be expressed as the summation of the mean wind speed V_{ave} and the related standard deviation σ:

$$V_{gust} = V_{ave} + k\sigma \tag{5.89}$$

The constant k depends on the averaging time of the mean and the gusts.

Using the more familiar parameter, that is turbulence intensity I, Equation (5.89) can be transformed into

$$V_{gust} = V_{ave}(1 + kI) \tag{5.90}$$

The calculation of k can be found in many works (e.g. see Reference [34]. For 3 s extreme gusts with a return period of 50 years, V_{ave} will be replaced by the 50-year extreme wind speed V_{50}, with I being the average turbulence intensity of the extreme wind speeds. For conservative reasons, the value of k can be approximated to be 3.4, 2.8 and 1.9 for hourly, 10-min and 1-min averaging wind speed time series respectively.

Here, a gust factor, G, can be defined as

$$G = \frac{V_{gust}}{V_{ave}} \tag{5.91}$$

Thus

$$G = 1 + kI \tag{5.92}$$

Typical values of gust factor G is 1.3–1.4 [33]. IEC 61400-1 [4] uses 1.4 to define the steady extreme wind model with a recurrence period of 50 years (see Table 4.1) and gives

$$V_{50,gust} = 1.4V_{ref} \tag{5.93}$$

under the assumption that constant yaw misalignment will be in the range of $\pm 15°$, and the averaging times for the extreme gust and mean speed are taken as 3 s and 10 min respectively.

The IEC 61400-1 [4] standard has defined extreme load cases based on a variety of extreme wind conditions besides the extreme wind speed and gust. These are the extreme operation gust (EOG), the extreme turbulence model (ETM), the extreme direction change (EDC), the extreme coherent gust with direction change (ECGDC) and the extreme wind shear (EWS). The extreme wind conditions expected to attack a wind turbine have to fall into certain categories defined by the IEC standard for the sake of turbine safety. From the wind engineering point of view, these extreme conditions can all be studied using the extreme value statistics presented in Section 5.6.2 and Section 5.6.3.

5.7 Summary

Statistics is one of the backbones of wind engineering owing to the stochastic nature of the wind. This chapter started with an introduction of the fundamentals in statistics for those who are unfamiliar with the subject, followed by the mathematical background for wind data sampling and statistics for mean and extreme wind speeds. Some statistical parameters for the wind, such as turbulence scales and wind frequency spectrums, are important for wind turbine load simulations, but are trivial for wind engineering for the purpose of wind resource assessment, and therefore were not included in this chapter. More statistical applications in wind engineering including wind data correlation and uncertainty analysis of production estimates will be introduced in the next two chapters, where the mathematical understanding from this chapter is still important.

Sample means and standard deviations of the wind speed and direction for a certain averaging interval (usually 10 minutes) are recorded in a wind data time series, the raw material for statistical analysis of the wind.

The frequency of the mean wind speed follows a Weibull distribution, which is determined by a shape parameter k and a scale parameter a. Five common parameter estimate methods for Weibull distributions are presented, from the graphical method (linear regression) to the universal maximum likelihood method in statistics to the power density method developed specially for wind energy applications.

The probability of extreme and rare wind events follows extreme value distributions, which are different from that for the mean wind speed. The generalised extreme value (GEV) distribution and the generalised Pareto distribution (GPD) are introduced as the statistical tools for estimating the 50-year extreme wind speeds. The Gumbel distribution (GEV Type I) is the most popular methodology with engineers because it is linear and graphical. The application of these extreme statistics requires the extreme wind speeds in use to be statistically independent and Poisson.

References

[1] Hwei, H. (2010) *Schaum's Outline of Probability, Random Variables, and Random Processes*, 2nd edition, McGraw-Hill.

[2] Wilks, S.S. (1962) *Mathematical Statistics*, Section 8.1, John Wiley & Sons, Ltd.

[3] Lundquist, J., Clifton, A. (2012) How Turbulence Can Impact Power Performance. *North American Windpower*. Available at: http://wndfo.net/D3086 (last accessed on 13 January 2014).

[4] IEC 61400-1 (Ed.3) (2005) Wind Turbine – Part 1: Design Requirements [S].

[5] Hansen, K.S., Larsen, G.C. (2006) De-trending of Turbulence Measurements. Presented at *OMEMES 2006*, Civitavecchia.

[6] Damaschke, M., Mönnich, K., Söker, H. (2007) *De-trending of Turbulence Measurements – Identification of Trends and Their Suppression*, DEWI, Germany.

[7] IEC 61400-12-1 (2006) Wind Turbines – Part 12-1: Power Performance Measurements of Electricity Producing Wind Turbines.

[8] Kaimal, J.C., Wyngaard, J.C., Izumi, Y., Coté, O.R. (1972) Spectral characteristics of surface layer turbulence. *Quarterly Journal of the Royal Meteorological Society*, **98**, 563–589.

[9] United States Environmental Protection Agency (2000) *Meteorological Monitoring Guidance for Regulatory Modeling Applications*. EPA-454/R-99-005.

[10] Mori, Y. (1986) Evaluation of Several Single-Pass Estimators of the Mean and the Standard Deviation of Wind Direction. *Journal of Climate and Applied Meteorology*, **25**, 1387–1397.

[11] Yamartino, R.J. (1984) A Comparison of Several 'Single-Pass' Estimators of the Standard Deviation of Wind Direction. *Journal of Climate and Applied Meteorology*, **23** (9), 1362–1366.

[12] Mardia, K.V. (1975) Statistics of Directional Data. *Journal of the Royal Statistical Society – Services B*, **37**, 349–393.

[13] http://www.renewablenrgsystems.com/ (last accessed on 15 January2014).

[14] Justus, C.G., Hargraves, W., Amir, M., Denise, R. (1978) Methods for Estimating Wind Speed Frequency Distribution. *Journal of Applied Meteorology*, **17**, 350–353.

[15] Langlois, R. (1991) Estimation of Weibull Parameters. *Journal of Material Science Letters*, **10**, 1049–1051.

[16] He, X., Fung, W.K. (1999) Method of Medians for Life Time Data with Weibull Models. *Statistics in Medicine*, **18**, 1993–2009.

[17] Akdag, S.A., Ali, D. (2009) A New Method to Estimate Weibull Parameters for Wind Energy Applications. *Energy Conversion Managment*, **50**, 1761–1766.

[18] Kantar, Y.M., Usta, I., Arik, I. (2013) The Method of Medians to Estimate Weibull Parameters for Wind Energy Applications. *Proceedings – Science and Engineering (2013)*, pp. 377–382.

[19] Zhang, L.F., Xie M., Tang, L.C. (2006) Robust Regression Using Probability Plots for Estimating the Weibull Shape Parameter. *Quality And Reliability Engineering International*, **22**, 905–917.

[20] Hyndman, R.J., Koehler, A.B. (2006) Another Look at Measures of Forecast Accuracy. *International Journal of Forecasting*, 679–688.

[21] Harris, R.I. (1996) Gumbel Re-visited – A New Look at Extreme Value Statistics Applied to Wind Speeds. *Journal of Wind Engineering and Industrial Aerodynamics*, **59**, 1–22.

[22] Søren, O. (2006) Extreme Winds in the Western North Pacific. Risø-R-1544(EN), Risø National Laboratory, Denmark.

[23] Kristensen, L., Casanova, M., Courtney, M., Troen, I. (1991) In Search of a Gust Definition. *Boundary-Layer Meteorology*, **55**, 91–107.

[24] Pielke Jr, R.A.C., Landsea, M., Mayfield, J. L., Pasch, R. (2005) Hurricanes and Global Warming. *Bulletin of the American Meteorological Society*, **86**, 1571–1575.

[25] Gumbel, E.J. (1958) *Statistics of Extremes*, Columbia University Press, New York.

[26] Coles, S. (2001) *An Introduction to Statistical Modeling of Extreme Values*, Springer Series in Statistics. Springer-Verlag, London.

[27] Ragan, P., Manuel, L. (2007) Statistical Extrapolation Methods for Estimating Wind Turbine Extreme Loads. *45th AIAA Aerospace Sciences Meeting and Exhibition*, Reno, Nevada.

[28] Makkonen, L. (2006) Plotting Positions in Extreme Value Analysis. *Journal of Applied Meteorology and Climate*, **45** (2), 334–340.

[29] Brabson, B.B. (2000) *Tests of the Generalized Pareto Distribution for Predicting Extreme Wind Speeds*, American Meteorological Society, USA.

[30] Pickands, J. (1975) Statistical Inference Using Extreme Order Statistics. *The Annals of Statistics*, **3**(1), 119–131.

[31] An, Y., Pandey, M.D. (2005) Technical Note – A Comparison of Methods of Extreme Wind Speed Estimation. *Journal of Wind Engineering and Industrial Aerodynamics*, **93**, 535–545.

[32] Walshaw, D., Anderson, C.W. (2000) A Model for Extreme Wind Gusts. *Applied Statistics*, **49**(4), 499–508.

[33] Emeis, S. (2013) *Wind Energy Meteorology*, Springer, New York.

6

Measure–Correlate–Predict

As stated in previous chapters, the wind is highly variable on short- and long-term scales. The expected lifetime of a wind farm project is usually 20 years or more, but for practical reasons (wind monitoring is expensive and time-consuming) the length of typical wind resource assessment campaigns for commercial wind power projects last anywhere from one to three years, with important decisions to be made often only after several months. One year of in situ wind data is usually taken as the minimum requirement for a reliable wind resource assessment, but it cannot capture year-to-year variations in average wind speed and other wind characteristics, as the measurement campaign may correspond to an untypically high or low period.

To address these issues, the measure–correlate–predict (MCP) methodology has been developed to expand the wind data series into a longer period of typically 10 years or more. This is done by establishing the statistical relation between site-specific short-term wind data (measure) and long-term reference wind data from nearby sites using the concurrent or overlapping data period (correlate), and then implementing the relation to the entire long-term data period (predict). MCP has become the industrial standard approach to address this issue. The level of correlation between the short-term and the reference dataset, the homogeneity of the reference dataset and the length of the overlapping period are the most important factors determining the quality of the prediction results [1]. Furthermore, since the power output of a wind turbine depends on the wind speed in a nonlinear way, it is not enough to have an estimation only of the mean (first-order moment) wind speed, but also of the variance (second-order moment) in particular. If possible, it is desirable to estimate the actual distribution of the wind speed values in each directional bin [2]. MCP can be considered as statistical wind flow modelling, which is used in combination with numerical wind flow modelling to achieve better wind resource assessment results.

There are various types of MCP methods, including regression, matrix methods and the neural networks capable of incorporating multiple sources of reference data. This chapter will introduce a few important MCP methods as well as the associated uncertainty in the predictions. The application of reanalysis data is also introduced in the chapter as they are most often used as reference datasets, even in the case where an meteorological station is not too far from the measuring site due to homogeneity issues [1].

Wind Resource Assessment and Micro-siting, Science and Engineering, First Edition. Matthew Huaiquan Zhang.
© 2015 John Wiley & Sons, Ltd. Published 2015 by John Wiley & Sons, Ltd.

6.1   Wind Data Correlation

Correlation analysis is the first step in any MCP procedure. It is to statistically establish the relation between the short-term measured wind data site, hereafter referred to as the target site, and the long-term reference data site, hereafter referred to as the reference site.

6.1.1   Correlation Coefficient

In statistics, correlation refers to statistical relationships of two random variables or two sets of data involving their probabilistic dependence. Correlations are useful because they can indicate a predictive relationship that can be exploited in practice. The more dependence the two sets of data, the better the predictive relationship (the Pearson product moment). The correlation coefficient, R, is the measure of dependence between the two random variables, X and Y, as defined by

$$R(X, Y) = \frac{\text{cov}(X, Y)}{\sigma_X \sigma_Y} \tag{6.1}$$

where the covariance cov is given by

$$\text{cov}(X, Y) = \frac{1}{n} \sum_{i=1}^{n} (X_i - \mu_X)(Y_i - \mu_Y) = E[(X_i - \mu_X)(Y_i - \mu_Y)] \tag{6.2}$$

and E denotes the expected (mean) value operator, with μ_X, μ_Y, σ_X and σ_Y the means and standard deviations of X and Y.

The correlation coefficient, R, ranges between -1 and $+1$. The value of $+1$ implies perfect positive correlation; that is if one variable moves, either up or down, the other variable will move in lockstep in the same direction. Alternatively, perfect negative correlation (-1) means the other variable will move in the opposite direction. If R is 0, the movements of the two variables will have no correlation.

In wind data correlation, the correlation coefficient commonly refers to R^2, in order to disregard the correlation direction. R^2 ranges from 0 to 1, with 1 being the perfect correlation. R^2 is a crucial parameter to indicate the level of correlation between the short-term and the reference datasets. However, it should be noted that higher values of R^2 do not always guarantee a more accurate prediction. The quality of prediction should be estimated by a similar parameter called the coefficient of determination (also refer to Section 5.5.6), which will be discussed later. Nevertheless, the quality of reference data can still be evaluated using Table 6.1.

The short-term and reference data series may be monitored in different time zones or averaged by different intervals. Thus, they have to be aligned in order to synchronise the signals before performing the MCP process and calculating the correlation coefficient.

6.1.2   Physical Interpretations of the Correlation

The likely dependence of the correlation between the target site and reference site can be explained qualitatively to some extent. Understanding the physical processes that may affect the level of correlation can also help us to choose the best possible reference sites.

Firstly, the reference site may be located somewhere with distinctive local topographic features (terrain, obstacle and roughness; see Chapter 2) than the target site. Even though a wind

Table 6.1 Quality of reference data

Correlation coefficient R^2	Quality of reference data
0.9 ~ 1.0	Very good
0.8 ~ 0.9	Good
0.7 ~ 0.8	Moderate
0.6 ~ 0.7	Poor
<0.5	Very poor

climate of larger scale is identical between the two sites, the wind flows may be severely altered by these local features, thus lowering the correlation between the two wind datasets. Therefore, due consideration must be given to local features when assessing these correlations.

Secondly, the distance between the two sites means that there may be time lags on the winds. In other words, it takes some time for the wind to propagate from one site to other, depending on the wind direction. A longer distance would foster weaker correlation in general. Due to the compounding effects of local features, however, one can only expect a weak relationship between distance and the level of correlation. Nevertheless, it may still be possible to improve the correlation by deliberately applying a small time lag between the target site and the reference site, especially when the two sites are in line with the prevailing wind direction. For large distances, the two sites may be exposed to different wind regimes overall, promising a weak correlation, especially for near-shore conditions and complex terrain.

The effects of atmospheric stability also play an important role for the correlation. Atmospheric stability influences vertical mixing of the wind. As a result, wind shear is generally greater at night than it is during the day when unstable vertical mixing is more prevalent. Since most reference sites used for MCP recording of wind speed near the ground level and target sites are taken near hub height, the correlation between the two datasets often varies with the time of day [3] and the time of the year. These diurnal and seasonal variations in the correlation relationship are of great importance for the MCP process.

If long-term historical wind data from surface meteorological stations are used as reference sites, the analyst has to check the homogeneity or consistency of the reference data, because inconsistency in reference data represents trends or discontinuities that are not real manifestations of the wind, thus giving a false interpretation of the correlation and wrong predictions of the long-term wind conditions. The inconsistency issue arises when the measurement height, location and instruments are changed and the surroundings develop (e.g. trees growing) over the years.

6.1.3 The Impact of Averaging Interval

When the site versus reference relationship is poor, the analyst may pre-process the data with longer averaging intervals of typically one day or one week. The resulting improvements on the level of correlation can be explained from two aspects. Firstly, longer averaging intervals can smooth the diurnal variations in atmospheric stability. However, the more significant impact of averaging interval is fundamentally due to the distance between the two sites. When the two sites are nearby, short averaging intervals (10 minutes or 1 hour) should be able to simultaneously record the same wind process (e.g. weather fronts and meso-scale circulations).

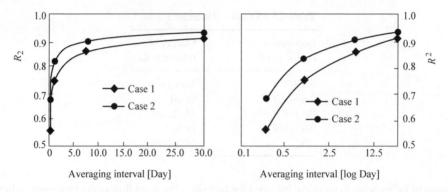

Figure 6.1 Examples of correlation coefficient plotted against the averaging interval. The four data points on each line represent averaging intervals of 6 hours, 1 day, 7 days and 30 days respectively

When the two sites are further apart, say 50 km to 100 km, the time scale of the general wind regime is of 6 hours or more and only an averaging interval that is longer than 6 hours should significantly improve the correlation.

Figure 6.1 illustrates two examples of the effect of the averaging interval on the correlation coefficient, R^2, of two wind datasets. As shown in the figure, the correlation coefficient improves very rapidly with the averaging interval when it is shorter than 1 day. Improvements are still noticeable from 1 day to 7 days, but generally diminish after 7 days; 7 days loosely correspond to the time scale of the passing of typical weather systems (e.g. cyclones and anticyclones) in middle latitudes.

Consequently, it seems desirable to use longer averaging intervals for MCP, but the improvements in correlation can be counteracted by the loss of energy content in the wind due to the fact that wind power density (WPD) is proportional to cubic wind speed. Three wind data time series from distinctive wind regimes in China are taken as examples of the relationship between the averaging interval and wind power density, as summarised in Figure 6.2. It is

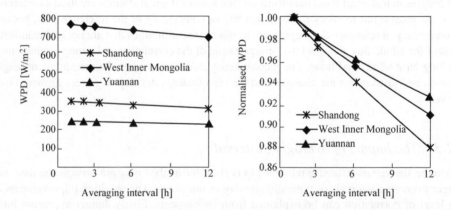

Figure 6.2 Wind power density (WPD) of three wind data series of distinctive wind climates (Shandong, Inner Mongolia and Yunnan, China) plotted against the averaging interval. The data points on each line represent averaging intervals of 1, 2, 3, 6 and 12 hours respectively

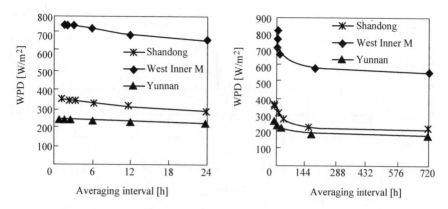

Figure 6.3   WPD of the same wind data series as Figure 6.2 plotted against the averaging interval of wider ranges (left: 1 h to 20 h; right: 10 days to 30 days)

shown in Figure 6.2 that the wind power density for all of the three wind data series decreases linearly with the averaging time, and around a 10% loss on wind power density can be expected when the averaging time decreases from 1 to 12 hours. The linear trend is only supposed to be maintained below 24 hours though, as shown in Figure 6.3 (left), after which the rate of decrease seems to have calmed down dramatically, as shown in Figure 6.3 (right).

Therefore, it may be possible to improve the MCP process in terms of predicting the energy content of the wind using the linear trend presented in Figure 6.2. As a rule of thumb, the optimal averaging time interval should be roughly equal to the size of typical weather disturbances divided by their average rates of motion.

6.2   Wind Data Regression and Prediction

6.2.1   Regression Equation and Residual

Regression analysis is a statistical process for estimating the relationships among variables. The estimation target is a transfer function called the regression function. In terms of wind data regression, let Y_i denote the target site wind speeds and X_i denote the concurrent reference site wind speeds; then the regression function is given by

$$Y_i = f(X_i) + e_i \tag{6.3}$$

and the predicted value of Y is given by

$$\hat{Y}_i = f(X_i) \tag{6.4}$$

where $f(X)$ is the regression model/equation (also referred to as the transfer function), which relates Y to a function of X, and e is the residual (or fitting error) representing the random error that is not explained by the regression equation.

A comparison between the wind speed time series of a target site and a reference site is illustrated in Figure 6.4. Even though the wind speeds of the two sites have significant discrepancies, they generally follow the same wave trend. Thus, establishing a mathematical relation between them should be possible.

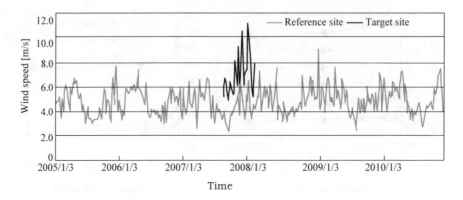

Figure 6.4   Wind speed time series of target and reference site

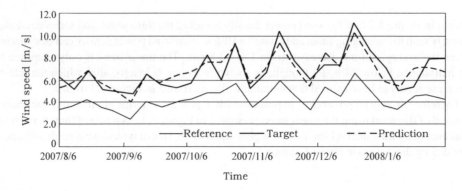

Figure 6.5   Wind speed prediction (interpolation) using the simple linear regression equation

Predictions can be made once the regression equation is determined. A prediction within the range of values in the dataset used for model-fitting or regression is known informally as an interpolation. A prediction outside this range of the data is known as an extrapolation.

Figure 6.5 shows the predictions using a simple linear regression technique for the concurrent data period in Figure 6.4. The differences between the predicted values and the actual values of the target site (prediction errors) are expressed by the residual e. If the regression equation predicts the actual value of Y well, both the sum and the mean of the residuals should be equal to zero (unbiased), and the distribution of e should follow a normal distribution, as expressed by

$$e \sim N(0, \sigma^2) \tag{6.5}$$

The analysis of the MCP performance is predominantly based on the statistical quantities of the residual e, for example its mean, variance, skewness and kurtosis.

Long-term wind variations at the target site are predicted by applying the regression equation to the entire period of the reference site (extrapolation) (see Figure 6.6). Performing an extrapolation relies strongly on the regression assumptions. The further the extrapolation goes outside the data, the more room there is for the model to fail due to differences between the assumptions

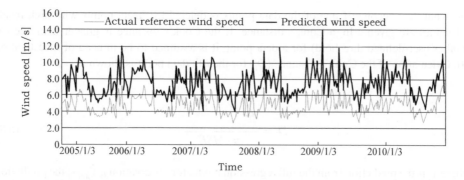

Figure 6.6   Wind speed prediction (extrapolation) using the simple linear regression equation

and the sample data or the true values. It is generally advised that an extrapolation should be performed in conjunction with validating the accuracy of a regression fit and analysing the associated uncertainty.

In an MCP process, the form of the regression equation is usually presumed, be it linear or parabolic or otherwise. The parameters of the presumed regression equation are usually estimated by minimising the sum of squared errors (or residuals), which is the standard approach, called the least-squares regression. The sum of squared errors (SSR) can be given by

$$SSR = \sum_{i=1}^{n} e_i^2 = \sum_{i=1}^{n} (Y_i - f(X_i))^2 \tag{6.6}$$

Instead of minimizing the sum of squared errors, a method that tries to minimise the sum of the absolute errors may be implemented as well. This results in a more robust regression model, which is less sensitive to outliers. This method is called the least absolute error (or deviation) regression [4] and can be given by

$$LAE = |Y_i - f(X_i)| \tag{6.7}$$

Standard regression assumes that the target site wind speed occurs and that errors only happen in the reference site. Orthogonal regression, where the distance between the measurements and the fitted line is minimized, is believed to be more suitable than a standard least-squares fit if both sets of measurements have an associated error [2, 5].

6.2.2   Data Validation

Outliners in the data set may not be technically valid and should not be included in the regression operation. If the outliners are due to errors, the resulting regression equation will be biased [6]. It is always possible to validate the data series graphically, normally by plotting the series against time. A change or defect in the measuring equipment may result in a different wind speed and can be noticed in the graph. The reduction of errors is possible if inconsistency in the time period is found and the predictions are confined to the consistent time period [7].

Statistically, Cook's distance can be used to detect and quantify outliers in a dataset. It is an influence measure based on the difference between the regression parameter estimates and what they become if one of the data points is deleted [8]. Data points with high residual values are given a high Cook's distance. If the Cook's distance is greater than one (or another user-defined threshold) for a data point, it is assumed to be in need of investigation [9]. Calculation of the Cook's distance (D_i) for the i-th data point is shown as

$$D_i = \frac{\sum_{j=1}^{n} (\hat{Y}_j - \hat{Y}_{j(i)})^2}{p \cdot MSE} \tag{6.8}$$

where $\hat{Y}_j$ is the prediction from the full regression model for observation j, $\hat{Y}_{j(i)}$ is the prediction for observation j from a refitted regression model in which observation i has been omitted, MSE is the mean square error of the regression model and p is the number of fitted parameters in the model.

The data points with a D_i value higher than the threshold cannot automatically be omitted, but should be treated as an alert for investigation. It should be noted that omitting outliners may improve the correlation, but important information about the data can be lost in the process as well.

6.2.3   Data Resampling

Data resampling consists of probabilistic methods generating 'artificial' time series for long-term corrected wind distribution. It is of interest to validate the equation given by a regression model [6].

In statistics, bootstrapping [10] is a method for assigning measures of accuracy (defined in terms of bias, variance, confidence interval, prediction error or some other such measure) to sample estimates, allowing estimation of the sampling distribution of almost any statistic using only very simple methods and a very small sampling size. It is therefore very useful when there are only a few data points available (a common case for wind data MCP) for the correct construction of the distribution of the entire population. In the case where a set of observations can be assumed to be from an independent and identically distributed population, this can be implemented by constructing a number of resamples of the observed dataset (and of equal size to the observed dataset), each of which is obtained by random sampling with replacement from the original dataset.

The Jack-knife is an alternative data resampling technique. It is less complex but more robust than bootstrapping and is capable of giving consistent results, even when the distribution is tailed. This is why the Jack-knife is usually recommended for wind energy applications. The Jack-knife method requires a certain smoothness level of the dataset [6].

Numerically, data resampling is based on the Monte Carlo simulation [11], which is a computerised mathematical technique used to approximate the probability of certain outcomes by running multiple trial runs called simulations, using random variables, allowing the decision makers to see the possible outcomes of your decisions and assess the impact of risk. It is most useful when it is difficult or impossible to obtain a closed-form expression or it is infeasible to apply a deterministic algorithm.

6.3   MCP Methodology for Wind Energy

A wide variety of MCP methodologies for wind engineering have been developed. The most commonly applied categories are based on regression techniques, such as linear regression, the variance ratio method [12], the support vector regression [13] and the Weibull regression (bivariate joint distribution) [2]. Other methods go beyond the regression approach, such as the Mortimer method [14], which retains some of the correlation structure between the two sites in the model, the Weibull scale method [15] and the wind index method [15]. Artificial neural networks (ANNs) [16], which treat the correlation as a complex system have also been implemented in recent years.

Apart from different methods used to fit the regression parameters (correlation techniques), another distinction of these MCP approaches refers to the treatment of wind direction [2]. Generally, the wind speed observations are grouped or binned by the wind direction at the reference site and/or other factors and then separate parameters are fitted for each bin. A comprehensive MCP approach should be able to give correct predictions of mean wind speed, wind speed distribution, annual energy production at the target site and wind direction distribution [12].

These MCP approaches typically used in the wind energy industry have been reviewed in many reports including Anderson (2004) [5], Rogers *et al.* (2005) [12], Sheppard (2009) [16] and Thøgersen *et al.* (2007) [15]. Their hybrids are also tested in Zhang *et al.* (2012) [17]. This chapter will attempt to elaborate on six representative MCP methods. Given the scope of this book, support vector regression and Weibull regression will not be included here.

6.3.1   Linear Regression

The conventional leastsquares linear regression is simple and well understood. It remains popular among many other more sophisticated MCP models partly because these models do not always perform substantially better [5, 18]. The simple linear regression MCP model for wind engineering was put forward by Derrick [19, 20], given by the form of

$$\hat{Y} = f(X) = aX + b \tag{6.9}$$

or logarithmically as

$$\log \hat{Y} = a \log X + b \tag{6.10}$$

where a and b are the slope and offset determined from linear regression.

Applying the simple linear regression to the concurrent period in Figure 6.4, the regression equation can be achieved, as shown by the following equation and Figure 6.7:

$$\hat{Y}_i = 1.5132X_i + 0.2292 \tag{6.11}$$

Equation (6.11) is used to make the predictions shown in Figures 6.5 and 6.6. In practice, wind data are binned into directional sectors at the reference site (usually twelve 30-degree sectors) and then separate parameters are calculated for each sector. If negative wind speeds are predicted, then they are set to zero. Omitting low wind speeds (<3–4 m/s) in the determination of correlation can improve energy capture estimates, but on the other hand can result in incorrect estimates of the overall mean wind speed [12].

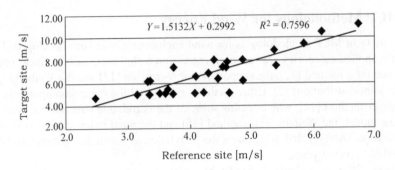

Figure 6.7   Simple linear regression of concurrent wind speeds in Figure 6.4

Wind direction is treated rather rudimentarily in the simple linear regression method. To accommodate direction differences at the two sites, Nielsen *et al.* [21] proposed a two-dimensional linear regression in orthogonal coordinates of the form:

$$\begin{bmatrix} \hat{Y}_1 \\ \hat{Y}_2 \end{bmatrix} = \begin{bmatrix} a_1 \\ a_2 \end{bmatrix} + \begin{bmatrix} b_{11} b_{12} \\ b_{21} b_{22} \end{bmatrix} \begin{bmatrix} X_1 \\ X_2 \end{bmatrix} \tag{6.12}$$

The final predicted wind speeds at the target site are

$$\hat{Y} = \sqrt{\hat{Y}_1^2 + \hat{Y}_2^2} \tag{6.13}$$

Instead of assuming that the wind direction at the target site is the same as that of the reference site, the two-dimensional linear regression method tries to predict the wind direction distribution at the target site as part of predicting the wind speed, resulting in a better prediction of the target site wind directions [12]. Rogers *et al.* (2005) [12] have pointed out, however, that this approach consistently underpredicts the wind speed due to the smaller variance of the predicted data compared to the observed data and the magnitude of the bias primarily depends on the coefficient of variation of the wind components and how well the model fits the data for the two components.

Woods and Watson [22] apply a binned method that is better at dealing with wind direction issues. They bin the data by direction sectors at both the reference and target sites, and then use linear regression to relate wind speeds in each bin and the matrix of frequencies of occurrence to determine the final wind speed and direction distributions. A similar binning idea has been adopted by many MCP methods such as the Mortimer and matrix methods to be introduced later.

In a broader sense, linear regression can be generalised to nonlinear methods by assuming second- or higher-order polynomial relationships between the reference and the target sites, but they cannot promise a better long-term prediction of the wind resource without proper supports from an underlying statistical or physical model [2].

6.3.2   Variance Ratio Method

With the simple linear regression, the standard deviation (variance) of the predicted wind speeds about the mean is generally smaller than that of the originally observed wind speeds because the correlation coefficient, R^2, is always smaller than 1 in practice. This can result in biased predictions of wind speed distributions. This is detrimental because the mean power production of a given wind turbine will depend on measures of wind speed fluctuations. Rogers *et al.* (2005) [12] attack this issue by suggesting that the variance ratio method is based on a simple linear regression. This method works really well while keeping the simplicity of linear regression.

The variance ratio model forces the variance of the predicted wind speeds to be equal to that of the observed ones, that is

$$\sigma^2(\hat{Y}) = \sigma^2(aX + b) = a^2\sigma^2(X) \tag{6.14}$$

Setting

$$a^2 = \frac{\sigma^2(\hat{Y})}{\sigma^2(Y)} \tag{6.15}$$

and ensuring that

$$\sigma^2(\hat{Y}) = \sigma^2(Y) \tag{6.16}$$

the predicted values are expected to have the same overall mean and variance as the observed values, and thus the variance ratio model is given by

$$\hat{Y} = \left(\frac{\sigma_Y}{\sigma_X}\right)X + \left[\mu_Y - \left(\frac{\sigma_Y}{\sigma_X}\right)\mu_X\right] \tag{6.17}$$

where μ_X, μ_Y, σ_X and σ_Y are the means and standard deviations of the two concurrent datasets. As concluded by Rogers *et al.*, the variance ratio method appears to give an unbiased estimate of the mean wind speed, Weibull shape parameter k, and capacity factor, and all with a very low standard deviation over all of the datasets tested in the report [12].

6.3.3 Weibull Scale Method

The Weibull s,cale method [15] is an empirical method with a very simple and radical presumption. It allows linear manipulation directly on the Weibull parameters (a, k) and the frequency distribution. This method should be done with caution on locations with significant non-Weibull distributions and when the modification of Weibull parameters and frequency needed is very large. Moreover, the method only works well when only small corrections on directional distribution are needed.

The Weibull scale method presumes that the relationship between the Weibull distribution parameters should follow the general relations:

$$\frac{k_{target}^{long}}{k_{ref}^{long}} = \frac{k_{target}^{short}}{k_{ref}^{short}} \tag{6.18}$$

and

$$\frac{a_{target}^{long}}{a_{ref}^{long}} = \frac{a_{target}^{short}}{a_{ref}^{short}} \tag{6.19}$$

where the superscript *long* and *short* represent long-term and short-term concurrent periods of the *target* and *reference* datasets respectively.

The relationship of frequency distributions follows the same structure of Equations (6.18) and (6.19), but the modified long-term frequency distribution must be normalised to 100% of n sectors under consideration, that is

$$f_{target,i}^{long} = \frac{\dfrac{f_{target,i}^{short}}{f_{ref,i}^{short}} f_{ref,i}^{long}}{\displaystyle\sum_{i=1}^{n} \left(\frac{f_{target,i}^{short}}{f_{ref,i}^{short}} f_{ref,i}^{long} \right)} \tag{6.20}$$

This method requires look-up in the appropriate Weibull distributions, calculating the correction table and then doing the linear scaling operations to get the long-term distribution for the target site.

6.3.4 Mortimer Method

In the Mortimer method [14], concurrent data at the two sites are binned by the wind direction sector and wind speed at the reference site. The ratios of the concurrent target and reference wind speeds are calculated for each bin, producing two matrices: a matrix of the average of the ratios, r, and a matrix of the standard deviation of the ratios, e, in each bin. The prediction equation is given by

$$\hat{Y} = (r + e)X \tag{6.21}$$

where e is a random variable with a triangular distribution of a prescribed mean and width (standard deviation).

The Mortimer method is not based on regression techniques per se and therefore does not assume the residuals to be normally distributed. It was suggested by Mortimer that the method may predict extreme wind speeds better than linear regression. Special treatments may be required when there are not enough data points to determine the average and standard deviations of the ratios in a given bin. The Mortimer method seems to give a biased estimate for the Weibull shape parameter k [12].

6.3.5 WindPRO Matrix Method

The general matrix method is based on a joint probability distribution. The matrix method in WindPRO [15] bins the wind data of both sites by a two-dimensional matrix made of wind speed bins (1 m/s resolution) and direction bins (1 degree resolution, enlargeable to 30 degrees), to calculate the changes in wind speed (speed-up) and wind direction (wind veer) in each bin. The equation transferring the wind speeds and directions from the reference site to the target site is modelled through a joint distribution between the two variables: wind speed-up, Δv, and wind veer, $\Delta \theta$, given in the form of

$$\Delta v = v_{site} - v_{ref}$$

$$\Delta \theta = \theta_{site} - \theta_{ref} \tag{6.22}$$

The result from this binning is a set of joint sample distributions of wind veer and wind speed-up, conditioned on the wind speed and the wind direction on the reference site (see Figures 6.8 and 6.9), as

$$f(\Delta v, \Delta \theta) \qquad (6.23)$$

These joint distributions, which are stipulated to hold regardless of the time frame considered, are represented either directly by the binned sample distributions using a bootstrap resampling technique or alternatively by a bivariate normal distribution when no (or only a few) sample data are available in a particular bin. Since most long-term corrected samples are typically based on the first approach, the influence of the normal distribution assumption is limited.

The parameters of the joint normal distribution (mean, standard deviation and correlation) are modelled through polynomials of any user-defined order. These parameters are modelled as 'slices' of polynomial surfaces:

$$P|(v_{ref}, \theta_{ref}) = \sum_{i=0}^{n} a_i \theta_{ref} v_{ref}^i \qquad (6.24)$$

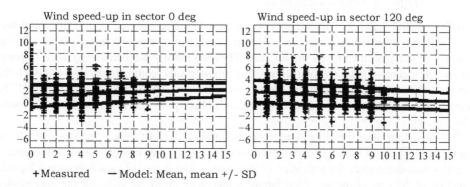

Figure 6.8   Sample data and first-order model for the wind speed-up (x = wind at reference and y = at speed-up) [15] *Source*: EMD International A/S

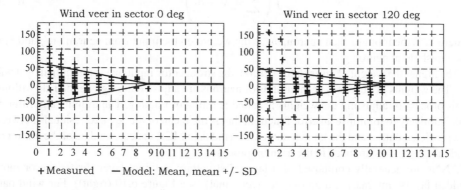

Figure 6.9   Sample data and first-order model for the wind speed-up (x = wind at reference and y = wind veer) [15] *Source*: EMD International A/S

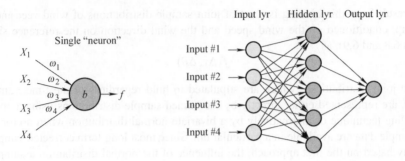

Figure 6.10   Illustration of a single neuron (left) and the layered structure of a feedforward ANN (right)

where P denotes the sample statistical moment considered, n is the order of the polynomial and a_i are the polynomial coefficients (which are also functions of θ_{ref}).

6.3.6   Artificial Neural Networks

Artificial neural networks (ANNs) are computational models inspired by animal central nervous systems, in particular the brain, which are capable of machine learning and recognising patterns in noisy or otherwise complex data. They are usually presented as systems of interconnected 'neurons' that can compute values from inputs by feeding information through the network [23]. Hundreds of different models considered as ANNs have been developed. The application of ANNs to wind resource MCP has also been proposed and tested in many reports [16, 18, 24].

The modelling of artificial neurons basically consists of inputs (like synapses), which are multiplied by weights (strength of the respective signals) and then computed by a mathematical function that determines the activation of the neuron. The output of each neuron is a logistic function of the sum of the weighted inputs minus some threshold value for the neuron (see Figure 6.10 (left)). The nonlinear output function of the neuron j can be expressed as [16]

$$f\left(\sum_j \omega_{ij} X_i - b_j\right) \tag{6.25}$$

where X_i is the input to each neuron, ω_{ij} is the weight on the input from neuron i to j and b_j is the threshold at neuron j.

The higher the weight of an artificial neuron, the stronger the input that is multiplied by it will be. The weights of hundreds or thousands of neurons are adjusted through defined algorithms. This process of adjusting the weights is called learning or training. The training algorithm is run iteratively to minimise an error function, that is the mean of the squared residuals between the predicted output and the known output. In wind data MCP, concurrent data of the reference and the target sites are used in the training process.

ANNs are generally composed of a layered structure: an input layer (first), one or more hidden layers (internal) and an output layer (final) (see Figure 6.10 (right)). For wind data MCP, the input layer would be the data from the reference site(s), while the final layer is the wind speed at the target site. Therefore, an ANN is developed by defining the interconnection pattern between different layers of neurons, the learning process for updating the weights of the interconnections and the activation function that converts a neuron's weighted input to output [17].

The advantages of neural networks in wind resource MCP are their ability to (1) recognise nonlinear relationships, (2) take past data (rather than simultaneous data pairs only) into consideration and (3) use more than one long-term data source [24], but they are generally difficult to implement.

6.4 MCP Uncertainty

MCP methods provide a single estimate of the wind resource at the target site. An estimate of the variance of that prediction is important for the analyst to describe the level of confidence in the accuracy of the results.

6.4.1 Reducing MCP Uncertainty

Theoretically, the possible sources of MCP uncertainty can originate from: (1) inconsistency and inaccuracy of reference data, (2) the ability of the chosen MCP model correctly modelling the relationship between the two datasets, (3) the length of the overlapping period to reflect the variability at time scales greater than the concurrent data length and (4) whether the relationship in the overlap period is representative of the long-term relationship (wind speeds, directions) [25, 26]. However, there is no direct way to quantify these uncertainties and the information available for determining the uncertainty is also very limited. Trying to reduce the associated uncertainty is certainly the sensible first option.

MCP uncertainty seems to have more to do with the data than the choice of the MCP method [18], provided that the correlation coefficient is higher than 80%. Thus, data consistency is often the most influential factor for MCP uncertainty. If possible, multiple nonsimilar reference datasets can be used for a consistency check and an indication of uncertainty. Quantifying inconsistency is usually difficult.

It is also helpful to assess from the reference data whether the wind regime in the overlapping period is representative of long-term windiness and whether the reference site is in a similar wind regime to the target site. It is also recommended to try several different MCP methods to check for sensitivity [25]. The length of a concurrent target reference period should be at least nine months, preferably a year or more in order to minimise uncertainty and capture seasonal variations in the correlation relationship.

6.4.2 Estimating MCP Uncertainty

We could intuitively conclude that a higher correlation coefficient implies lower uncertainty. Whilst this implication can be held to be true, quantification of the uncertainty using this correlation coefficient cannot be justified, mainly because the correlation of a concurrent period may not be representative for the long term and the nonrepresentativeness can be independent of the correlation in the concurrent period. The correlation is also affected by the range of wind speeds occurring [25].

As described earlier, the coefficient of determination (refer to Section 5.5.6) is more suitable than the correlation coefficient to indirectly describe the quality of MCP results. The coefficient of determination describes the sum of errors of the predicted dataset at the target site directly (post-MCP), whilst the correlation coefficient describes the sum of errors of the reference dataset with respect to the target site observations (pre-MCP). A low correlation coefficient may result in a relatively higher coefficient of determination, and vice versa, depending on the MCP process.

While acknowledging the incompetence of the correlation coefficient, some studies have given approximate relations between the correlation coefficient and the overall uncertainty in the predicted long-term mean wind speed. These relations usually account for the uncertainty inferred by the length of the reference period. For instance, Brower (2012) [27] presents the following approximate equation, assuming consistent reference data and a normally distributed annual mean wind speed:

$$\sigma = \sigma_{ann} \sqrt{\frac{R^2}{N_{ref}} + \frac{(1 - R^2)}{N_{target}}} \tag{6.26}$$

where σ_{ann} is the standard deviations of the annual mean wind speeds, N_{ref} is the number of years of the reference dataset and N_{target} that of the concurrent dataset. Equation (6.26) is not applicable if $N_{target} < 1$ is due to seasonal variations.

It is also pointed out in Brower (2012) [27] that the typical range of the standard deviation of annual mean wind speeds in Northern America is 3–6%. In Northern Europe and other areas in mid-latitude, the recommended value is usually 6%, for conservative reasons. Equation (6.26) is suspected to underestimate the uncertainty depending on the correlation coefficient, and thus is only advisable for a correlation coefficient larger than 80%.

To directly estimate MCP uncertainty, the concurrent data are usually broken into shorter periods (or by dropping out a nonoverlapping subsection of the concurrent data each time) and each shorter period is used to predict the long-term average or other statistic (e.g. Weibull parameters). The standard uncertainty of the MCP process is indicated by the standard deviation of the errors (root mean square errors, RMSEs) of multiple predictions.

For linear regression, Derrick [19, 20] has proposed an approach for estimating the uncertainty of MCP wind speed predictions using the variances and covariance of the slope and offset. However, this approach is confined to linear regression and wind speed results. It also significantly underestimates MCP uncertainty when the datasets are not independent and serially correlated, leading to unjustified confidence in the results [26].

The Jack-knife estimate of variance for estimating the uncertainty of the MCP predictions has been implemented by Rogers et al. (2006) [26]. They remarked that the Jack-knife method is capable of estimating the uncertainty of the results of any MCP model and the uncertainty of derivatives of the MCP wind speed results, such as Weibull parameters or wind turbine capacity factors. Rogers et al. (2006) [26] have also concluded that on average the Jack-knife method still underestimates the uncertainty of the mean wind speed and Weibull k predictions by about 38% and 18% respectively, and therefore further investigations on the Jack-knife method and data behaviour are necessary.

6.4.3 Overlapping Period

Rogers et al. (2006) [26] presents the relationship between MCP uncertainty and concurrent data length using normalised (site mean wind speed) standard deviations of MCP predictions

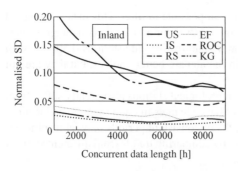 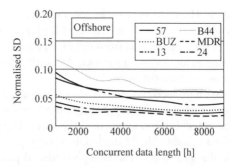

Figure 6.11   Measured normalised standard deviation (standard deviation/mean wind speed) of the mean wind speed for six inland and six offshore pairs of data [26]

of six inland and six offshore pairs of sites, as shown in Figure 6.11. The graphs show that in general the inland pairs of sites have a wider range of variability of the MCP estimates than the offshore sites.

Generally, MCP uncertainty decreases as the length of the concurrent period increases. The normalised standard deviation of MCP estimates may be as high as 20% of mean wind speed when using 1000 hours (about 1.5 months) of concurrent data. Lackner *et al.* [28] proposed a 'round-robin' method that was said to be able to improve the results. As we can see from Figure 6.11, the uncertainty level only starts to level off at about 7000 hours, which corresponds to about 9 months. Therefore, 9 months of concurrent data are usually taken as the minimum requirement for a confident MCP prediction.

Furthermore, this inverse trend is obvious up to 8–12 months, but may decrease slightly for around 1.5 years, and then increases again to 2 years, mainly due to seasonal variations in the correlation relationship [25]. Thus, a full year of concurrent data length is optimal. In the case of the concurrent data length greater than 1 year, between 1 and 2 years, it is strongly recommended to perform multiple MCP calculations each of exactly 1 year in order to get an indication of the prediction uncertainty using the standard deviation of the errors of the results.

6.5   Sources of Reference Data

There are generally two categories of reference data sources typically used in wind resource MCP. They are surface meteorological stations and modelled datasets, mainly competing on consistency, noise and availability.

6.5.1   Meteorological Stations

Surface meteorological stations are the conventional source of long-term reference wind data. A weather monitoring system is well established in most countries. Therefore, it is highly likely to find at least one representative meteorological station near a wind park project area. Instead of true average values, the data from surface meteorological stations is more likely to be averaged over very short intervals and only recorded hourly.

The drawbacks of the data from surface meteorological stations are clear. They are heavily influenced by local topographical features, resulting in poor correlation with target site data. The noise induced by local features is usually impossible to filter. This means that good correlation can only be expected for longer average intervals or at least daily average values. For this reason, data from tall meteorological towers and rawinsonde stations (balloons with sensors and radio transmitters measuring wind in the air) are better choices, but they are not available in most places.

As introduced previously, data from surface meteorological stations also greatly suffer from issues of inconsistency, because most stations will have changes in equipment, mast location, height and surroundings over the years. The lengths of consistent periods may be insufficient for a valuable MCP calculation. Verifying the data by comparing trends from different stations in the same region can be a good idea, if possible.

Generally speaking, surface meteorological station data are challenging to work with and other sources, if available, are usually preferred.

6.5.2   Reanalysis Data

Atmospheric reanalysis uses a constant data assimilation system to ingest observational data from a wide range of sources, such as surface weather stations, weather balloons, airport reports, commercial aircraft and satellite measurements. The ingested data are then used as input into a numerical weather prediction model to generate a description of the state of the atmosphere on a uniform horizontal grid and at uniformly spaced time instants.

Liléo et al. (2013) [29] has reviewed a variety of state-of-the-art long-term reanalysis datasets available for use in the long-term correction of wind measurements, including NOAA-CIRES (2 × 2 degrees horizontal resolution and 6 hours temporal resolution), which only assimilates surface observations of synoptic pressure into the global atmospheric model; MERRA (1/2 × 1/3 degrees and 6 hours) produced by the Global Modelling and Assimilation Office (GMAO) of the NASA Goddard Space Flight Centre; ERA-Interim (0.75 × 0.75 degrees and 6 hours) and ERA-40 (1.125 × 1.125 degrees and 6 hours) by the European Centre for Medium-Range Weather Forecasts (ECMWF); JRA-25 (1.25 × 1.25 degrees and 6 hours) and JRA-55, jointly produced by the Japan Meteorological Agency (JMA) and the Central Research Institute of Electric Power Industry (CRIEPI), the most commonly used NCEP/NCAR dataset. This section will choose the NCEP/NCAR as an example for using reanalysis data as the reference.

The National Centre for Atmospheric Research (NCAR) and the National Centre for Environmental Prediction (NCEP) of the United States collaborate in the production of the reanalysis data, often referred to as NCAR data for short. NCAR data comprise the second round of analysis (reanalysis) of observations collected worldwide, the first round being the production of real-time forecasts. NCAR data are available globally on a 210-kilometre grid (2 degrees in latitude and longitude) at 28 levels above ground in six-hour time intervals from 1948 to the present. The data are updated periodically and describe a wide variety of atmospheric parameters including temperature, pressure winds, moisture, etc.

The data are available online for the public to download[1], which makes it very convenient for MCP. Another unparalleled advantage of the reanalysis data is that they cover a period of more than 50 years of historical data, whereas most other reference data sources do not extend back that far. It should be noted, however, that a longer historical period does not necessarily result

---

[1] http://www.esrl.noaa.gov/psd/data/gridded/data.ncep.reanalysis.surface.html.

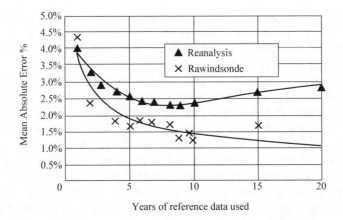

Figure 6.12   Dependence of the standard error of a 10-year predicted mean speed on the number of years of historical data used in the prediction, produced by sampling reanalysis data at every grid point in the United States. The result for a sample of rawinsonde stations is shown for comparison [30] *Source*: AWS Truepower, LLC

in a better MCP prediction (see Figure 6.12). The graphs in Figure 6.12 indicate that the MCP uncertainty using NCAR data as the reference drops quickly up to 8–9 years of data length, but rises after 10 years. This behaviour is caused primarily by inconsistency in the NCAR dataset. Even though scientists have tried to maintain the consistency of the reanalysis data through time by keeping the data assimilation system the same and frozen since 1995, the type, quality and quantity of observations put into the reanalysis system have changed dramatically over the decades [30]. Therefore, it generally makes little sense to use more than 10 years of reanalysis data for MCP unless the consistency of the data is validated.

Another positive attribute of NCAR data can be the fact that the NWP model in use takes little notes on micro-scale topographical influences, leading to less 'noisy' datasets compared to surface meteorological stations.

Unfortunately, the grid resolution of the global NCAR reanalysis data is relatively coarse. In regions of mountainous terrain or coasts or anywhere with a sharp wind gradient, the grid nodes may not be representative for the wind climate of the target site. In recent years, long-term historical wind data series generated by meso-scale NWP models (meso-scale reanalysis data) with much higher resolutions have become more available in the market, such as WRF FNL and WRF ERA-Interim, both produced by Kjeller Vindteknikk using the meso-scale model WRF [29].

6.6   Summary

It is critical to understand the long-term wind resource for the development, investment and operation of wind power plants. However, in practice, often only one year of onsite wind data, if not less, is available to make these important decisions. Therefore, there is a persistent urge in the industry to develop a reliable technique to extend the short-term wind measurements into a longer period of typically 10 years or more. This technique, based on statistic modelling, is called measure–correlate–predict (MCP). This chapter introduced the fundamentals of MCP methodology and six MCP methods typically used in wind resource engineering. These are linear regression, variance ratio, Weibull scale, Mortimer, WindPRO matrix and artificial neural network.

The level of correlation between the target and reference sites is the main issue of concern in MCP. It is physically influenced by their respective micro-scale flow features, the time it takes the same weather disturbance to synchronise between the two sites and seasonal and diurnal variations in atmospheric stability. On the other hand, the correlation has to be validated through omitting unrealistic outliners and removing false trends in the reference data. Thus, consistency of the reference data through time is of paramount importance for a reliable MCP analysis. It is often more important than the actual MCP methodology chosen.

The long-term prediction of the wind at the target site is based on the short-term overlapping data period between the target and the reference sites. It is clearly undeniable that the overlapping or concurrent period should be long enough to make statistical sense. Many studies have shown that a minimum of 8 to 9 months of the concurrent period is necessary for a stable and reliable analysis.

The analysis of associated uncertainty is indispensable for almost any statistical analysis. Without a proper understanding of the uncertainty, the predictions made by MCP cannot be trusted. After all, the purpose of MCP analysis in wind resource engineering is to reduce the uncertainty of long-term predictions. Unfortunately, however, there is no direct way to estimate MCP uncertainty. The correlation coefficient and coefficient of determination calculated from the concurrent data do not cover the entire long-term period, and thus can only be considered as an indirect indicator of the uncertainty.

The possible choices of reference data and their important pros and cons are introduced in the last section. Data consistency and representation of the wind climate at the target site are the qualities of reference data that should primarily be sought.

References

[1] Hanslian, D., Hosek, J. (2011) The Use of Measure–Correlate–Predict (MCP) Methods in Wind Energy Applications. *EMS Annual Meeting Abstracts*, **8**, EMS2011-739.

[2] Perea, A.R., Amezcua, J., Probst, O. (2011) Validation of Three New Measure–Correlate–Predict Models for the Long-Term Prospection of the Wind Resource. *Journal of Renewable and Sustainable Energy*, **3**, 023105.

[3] Oliver, A., Zarling, K. (2009) Time of Day Correlations for Improved Wind Speed Predictions. RES Report, AWEA, United States.

[4] Good, P.I., Hardin, J.W. (2006) *Common Errors in Statistics (*and How to Avoid Them*)*, 2nd edition, John Wiley & Sons, Inc., New Jersey.

[5] Anderson, M. (2004) A Review of MCP Techniques. Report No. 01327R00022 (Issue 3), Renewable Energy Systems, USA.

[6] Jonsson, C. (2010) Statistical Analysis of Wind Data Regarding Long-Term Correction. Uppsala University, Sweden.

[7] Siddabathini, P. (2008) Uncertainty in Long Term Predictions – A New Approach to Reduction. Master's Thesis, Aalborg University, Denmark.

[8] McDonald, B. (2002) A Teaching Note on Cook's Distance – A Guideline. *Res. Lett. Inf. Math. Sci.*, **3**, 127–128.

[9] Cook, R.D., Weisberg, S. (1982) *Residuals and Influence in Regression*, Chapman & Hall, New York.

[10] Efron, B., Tibshirani, R.J. (1993) *An Introduction to the Bootstrap, Monographs on Statistics and Applied Probability*, Chapman & Hall, Boca Raton, FL.

[11] Sørensen, J.D., Enevoldsen, I. (1996) Lecture Notes in Structural Reliability Theory. Aalborg University, Denmark.

[12] Rogers, A.L., Rogers, J.W., Manwell, J.F. (2005) Comparison of the Performance of Four Measure–Correlate–Predict Algorithms. *Journal of Wind Engineering and Industrial Aerodynamics*, **93**, 243.

[13] Mohandes, M.A., Halawani, T.O., Rehman, S., Hussain, A.A. (2004) Support Vector Machines for Wind Speed Prediction. *Renewable Energy*, **29**(6), 939–947.

[14] Mortimer, A.A. (1994) A New Correlation/Prediction Method for Potential Wind Farm Sites. *Proceedings of BWEA*, UK.

[15] Thøgersen, M.L., Motta, M., Sørensen, T., Neilsen, P. (2007) Measure–Correlate–Predict Methods: Case Studies and Software Implementation. EMD International A/S, Denmark.

[16] Sheppard, C.J.R. (2009) Analysis of the Measure–Correlate–Predict Methodology for Wind Resource Assessment. Master's Thesis, Humboldt State University, Arcata, CA.

[17] Zhang, J., Chowdhury, S., Messac, A., Castillo, L. (2012) A Hybrid Measure–Correlate–Predict Method for Wind Resource Assessment. *ASME 2012 6th International Conference on Energy Sustainability and 10th Fuel Cell Science, Engineering and Technology Conference*, San Diego, CA.

[18] Bass, J.H., Rebbeck, M., Landberg, L., Cabre, M., Hunter, A. (2000) An Improved Measure Correlate Predict Algorithm for the Prediction of the Long Term Wind Climate in Regions of Complex Environment. Joule Project Report JOR3-CT98-0295, EU.

[19] Derrick A. (1992) Development of the Measure–Correlate–Predict Strategy for Site Assessment. *Proceedings of BWEA, 1992*, UK.

[20] Derrick A. (1993) Development of the Measure–Correlate–Predict Strategy for Site Assessment. *Proceedings of EWEC, 1993*.

[21] Nielsen, M., Landberg, L., Mortensen, N.G., Barthelmie, R.J., Joensen, A. (2001) Application of Measure–Correlate–Predict Approach for Wind Resource Measurement. *Proceedings of EWEA, 2001*.

[22] Woods, J.C., Watson, S.J. (1997) A New Matrix Method of Predicting Long-Term Wind Roses with MCP. *Journal of Wind Engineering and Industrial Aerodynamics*, **66**, 85–94.

[23] Rojas, R. (1996) *Neural Networks: A Systematic Introduction*, Springer, Berlin.

[24] Albrecht, C., Klesitz, M. (2007) Long Term Correlation of Wind Measurements Using Neural Networks. *EWEC, 2007*.

[25] Faulkner, S. (2010) Quantifying the Uncertainties of Correlation–Prediction (MCP). Presentation at *NZ Wind Energy Conference, 2010*.

[26] Rogers, A.L., Rogers, J.W., Manwell, J.F. (2006) Uncertainties in Results of Measure–Correlate–Predict Analyses. *EWEC, 2006*.

[27] Brower, M.C. (2012) *Wind Resource Assessment: A Practical Guide to Developing a Wind Project*, 1st edition, John Wiley & Sons, Inc., USA.

[28] Lackner, M.A., Rogers, A.L., Manwell, J.F. (2008) The Round Robin Site Assessment Method: A New Approach to Wind Energy Site Assessment. *Renewable Energy*, **33**, 2019–2026.

[29] Liléo, S., Berge, E., Undheim, O., Klinkert, R., Bredesen, R.E., Vindteknikk, K. (2013) Long-Term Correction of Wind Measurements – State-of-the-Art, Guidelines and Future Work. Elforsk Report 13, p. 18.

[30] Brower, M.C. (2006) The Use of Reanalysis Data for Climate Adjustments. AWS Truewind Research Note #1, AWS Truepower, LLC.

7

Wind Park Production Estimate

The wind energy industry requires a correct estimate of how much energy a potential development wind park can produce in a lifetime of normally 20 years. Combined with an estimate of capital investment, the results can make or break the economics of wind park development. Quite different from traditional energy in terms of investment recovery, wind power plants do not exhaust any fuel for its operation and require very little operating costs. The economics of wind park development depends mainly on the initial investment and the estimated energy production. However, the stochastic nature of the wind and the accompanied uncertainties throughout the evaluation process create immense challenges for professionals in the field. While the expectations from the industry are understandable, they can be unrealistic sometimes as well. In general, a relatively conservative approach would be proposed when something uncertain is identified.

There are a number of software packages available on the market for wind resource assessment and wind park production estimates including WAsP [1], WindPRO (EMD International) [2], WindFarmer (Garrad Hassan) [3], WindFarm (Resoft) [4], openWind (AWS Truepower) [5], Meteodyn WT [6] and WindSim [7]. The fundamentals of these programmes have been introduced in Chapter 3. They all share many common features and capabilities, such as wind data analysis, generating wind statistics, handling geographical data, wind flow and wake modelling, layout optimisation, calculation energy production, etc. Despite their differences, the procedure of production estimates can be generalised as: (1) setting up geographical data of the project including height contours, roughness mapping, site boundaries, land use and restricted areas as detailed as possible, (2) analysing wind data and generating wind statistics, (3) turbine selection and layout design, (4) park simulation to calculate energy production and (5) analysing production loss and uncertainties.

This chapter will introduce a production estimate for a wind park project, not so much on the aspect of operating software tools but rather by focusing on correcting the estimated production and understanding its uncertainties in order to achieve more confident results for banking and financial profiling purposes.

7.1 Gross and Net AEP

The gross annual energy production (AEP) is the annual output of the wind park without wake and other losses. The gross AEP (kWh or MWh) for turbine k in the layout is the sum

Wind Resource Assessment and Micro-siting, Science and Engineering, First Edition. Matthew Huaiquan Zhang.
© 2015 John Wiley & Sons, Ltd. Published 2015 by John Wiley & Sons, Ltd.

of corresponding energy production in each direction sector and wind speed bins. It can be expressed as the sum of the following two-dimensional matrix:

$$AEP_k = 8760 \sum_{i}^{N_D} \sum_{j}^{N_U} f_{ijk} P_{ijk} \tag{7.1}$$

where N_D and N_U are the number of direction sectors and wind speed bins respectively, f_{ijk} is the fractional frequency of occurrence and P_{ijk} the energy production for direction sector i and wind speed bin j; 8760 denotes the number of hours in a year (24 hours a day, 365 days a year).

If the frequency of occurrence is derived from the overall Weibull distribution (see Chapter 5 of all direction sectors, Equation (7.1) can be simplified into the sum of a one-dimensional matrix:

$$AEP_k = 8760 \sum_{j}^{N_U} f_{jk} P_{jk} \tag{7.2}$$

The gross annual energy production calculated by wind resource assessment programmes usually has the wake loss included as it is the direct product of wind flow modelling. Other losses are estimated based on best practices and project features.

Net AEP is the gross AEP minus production losses. A good estimate of production losses should close the gap from theory to reality in terms of financial performance of the wind power project in prospect. There are six categories of production losses for a wind power project, which are usually expressed as a percentage of energy production. These are wake losses, availability losses, turbine performance losses, electrical losses, environmental losses and losses due to curtailments, all of which are very much site or turbine specific and can change over time.

These loss factors are not statistically independent, that is they may happen at the same time or individually. Direct addition of these losses would double count the overlapping portions. Therefore, they are combined by multiplying the efficiencies, which are defined as one minus the loss. The total production loss, L_{total}, is thus of the form:

$$L_{total} = 100\% - \prod_{i}(100\% - L_i) \tag{7.3}$$

where L_i denotes the production loss i as a percentage.

Typical total production loss of a wind power plant is around 20%, but may range from 10% to as high as 40% in some extreme cases. From an economic point of view, the discrepancy in each case is dramatic. A realistic estimate of each individual loss category is thus of vital importance for wind power financing. Reducing the losses is also the key to optimising a wind power plant.

7.1.1 Wake Losses

Wind turbine wake effects have been introduced in detail in Section 4.3. Reducing the production loss incurred by wake effects is the task of turbine layout optimisation. The wake losses for a wind power plant may range from 3% to 15%. Higher than 15% could be considered an unfeasible layout as such a high wake loss also means that the turbines are running under dangerous wake turbulence. For these reasons, wind turbines are rarely spaced closer than 5 rotor diameters onshore and 9 rotor diameters offshore in the prevailing wind direction and 3 and 5 rotor diameters in the crosswind direction for onshore and offshore wind parks respectively.

It is worth reiterating the effect of deep-array wakes in large scale wind parks where standard wake models tend to severely underestimate the wake loss. The deep-array wakes may occur in the future as the wind park may expand and other wind parks may be developed by other developers in adjacent areas. In this case, the wake loss can be time dependent through the lifetime of the wind power plant.

7.1.2 Availability Losses

Availability is the probability that a wind turbine (turbine availability) or a wind power plant (plant system availability) will operate without failure and within specified performance limits at any point in time, where the total time considered includes operating time, active repair time, administrative time and logistic time [8]. Turbine availability is a measure of the turbine only as it excludes wind farm system downtime caused by a failure or downtime of power grid or substation and force majeure [9]. Its definition is one of the main focuses of contractual warranty arrangements and negotiated between the wind park developer and the wind turbine manufacturer on a project by project basis before project construction. Plant system availability more closely represents the availability of the wind park as a whole regardless of the cause of downtime, except shutdowns due to normal operation conditions, such as cut-off at high wind speeds or cable unwind events, and thus should be the availability used in financial models of a wind power project [9]. Availability loss may be defined as one minus the system availability. In statistics, availability analysis involves the study of reliability and failure probabilities [8], which is outside the scope of this book.

An average overall plant availability loss of 5% is typically assumed, but it may vary between 2% to 10% (or even higher) from case to case depending on the confidence the analyst has for this turbine model and the manufacturer. The historical records and reputation of the turbine model and the manufacturer, service availability of the manufacturer in the market, technical maturity of the turbine model, etc., may all be components under consideration. The availability loss may also be site specific for a given turbine model. Wind conditions that tend to increase turbine loads, such as high turbulence intensity and high wind speed, should imply more energy loss as higher turbine loads can cause more downtime for the turbine.

Availability loss changes over time in a trough-like pattern. It normally drops significantly in the first year of the run-in period in operation and stabilises at a low level for the majority of the lifetime, before increasing again when the turbine approaches the end of its service. Therefore, it makes sense to assume a higher availability loss for the first one and the last three years of the turbine lifetime in the financial model.

7.1.3 Power Curve Performance

A power curve warranty is also an important part of the contract between the wind park project developer and the wind turbine manufacturer. Usually, the power curves advertised by manufacturers are tested and certified using stringent procedures defined by IEC standards [10, 11], keeping in mind the fact that false advertising is possible, but certification can only be given under well-defined wind conditions, which may depart from the wind conditions in reality.

Those wind conditions that may affect the power curve include high turbulence intensity, wind shear, inflow angle, turbulence length scale and air density. Furthermore, tilting and yawing of wind turbines, which may substantially reduce power in complex terrain, are not taken into account in power performance measurements [12]. Typical uncertainty of the measured power curve based on IEC standards is about 5% [13].

For a pitch-regulated turbine, high turbulence intensity tends to increase power production below rated wind speeds while decreasing power production at high wind speeds above rated conditions due to the concave upwards nature of the power curve. It has been reported [13] that an average of a 2% drop in nominal energy between turbulence intensity of 14% and 10% can be expected for a high wind speed site due to 'rounded knee' of the power curve and a 3% drop during periods of low turbulence intensity (<8%), which corresponds to stable atmospheric conditions. Power performance of wind turbines is more sensitive to wind shear (vertical wind profile) across the rotor at lower wind speeds where the effect can be of visible significance. A high inflow angle to the rotor seemingly reduces the power performance as well. It is also suggested that higher frequency turbulence might be expected to average out over the rotor disc whereas lower frequencies might have more coherence across the disc, thus intuitively leading to better energy conversion [14]. Air density varies with time of the day and the year, and so does the real power curve performance, as the power curve used in an energy production estimate is usually defined at the annual average or the standard ($1.225 \, kg/m^3$) air density (see Section 4.2.4). These conditions impacting the power curve are mainly influenced by atmospheric stability, terrain complexity and forestry.

Moreover, a wind turbine may not operate in optimal mechanical conditions, such as yaw misalignments, inaccurate blade pitching and control anemometer calibration errors.

It is reasonable to assume a 3–5% aggregated production loss due to power performance of wind turbines, depending on the turbine model itself, the wind conditions and even the proficiency of the turbine operator and the reputation of the manufacturer. After investigating many wind parks, Keir Harman from GL Garrad Hassan reported the results given in Table 7.1 at the EWEA Conference in 2012.

7.1.4   Environmental Losses

This category of production losses encompasses losses due to the aggregated effects of the environment, as in degradation of power performance and shutdowns triggered by environmental conditions outside the operating envelopes. The accumulation of damage

Table 7.1   Production loss due to turbine performance [13]

Loss category	Typical range of loss/gain	Median
1. Generic power curve performance loss (turbine model specific)	−5% to +3%	−1%
2. Mechanical suboptimal performance loss (operator specific)	−5% to 0%	−1%
3. Wind conditions – turbulence intensity, wind shear and flow inclination (site specific)	−5% to 1%	−1%

to the blades caused by icing, bugs, dirt and paint shedding may slowly reduce power production. Shutdowns can be triggered by too high or too low temperatures, lightening, icing, the accessibility to the site in bad weather for repairs and force majeure. Shutdowns due to icing build-ups can be especially significant in some regions. For example, in the mountainous Guizhou province in Southern China where high humidity is combined with sudden temperature drops at night, icing alone may prevent the turbines from producing any energy at all for over two consecutive months each winter.

These losses are mostly region-specific and may be evaluated sometimes from the data collected on site, including temperature records, lightening frequency maps, icing events observed in wind data series and observed on site, etc. Due to the difficulty of accessing the sites, especially remote offshore wind parks, often implies greater losses than normal. Depending on the region and the specific site, it is typical to assume 2–3% environmental losses.

7.1.5 Electrical Losses

Electrical losses in a wind power plant system originate from three major sections: the wind turbine system, transmission cables and substation system. Electrical losses inside a wind turbine generator vary with the electrical configuration. Permanent-magnet generators do not need to draw reactive power from the grid to excite the magnetic field. Variable-speed double-fed generators are able to generate power at a range of rotor speeds, whereas fixed-speed squirrel cage generators have to draw power from the grid to maintain functioning when the rotor speed is lower than a certain value. Converter and inverter losses may also be different from turbine to turbine. All of these features may lead to slightly different electrical losses inside the generator system.

Physically, though, electrical losses can be generalised into copper losses, iron losses, stray losses and consumption. Copper losses are resistance losses occurring in the winding coils in the generator, transformers and cables, etc. They are in a squared relationship with voltage. As the rated power of turbine generators and the tower height increases, the copper losses in the cables inside the tower become a major factor in terms of electrical losses [15]. In Vestas V90-2.0 MW and V112-3.0 MW, the low voltage/medium voltage power transformer is placed in the nacelle to achieve lower cable losses. Iron losses are mainly produced by the flux change in the generator and transformers and consist of eddy current losses and hysteresis losses. All losses are in general in a nonlinear relationship to the wind speed, making them difficult to determine. Taking iron loss as an example, when the wind speed increases, the generator's real power increases and the generator draws more reactive power and the internal voltage of the generator decreases. As a result, flux density and iron loss decrease [16].

Madariaga et al. investigated five alternatives of MW turbine configurations and reported electrical losses of 3.21% to 4.41% [15]. Compounding electrical losses at other sections of a wind power plant system, a total of 4% to 5% electrical losses on energy production seems justifiable for typical cases.

7.1.6 Curtailments

Curtailments occur when the turbines have to be down rated or even shut down for environmental reasons, like night-time noise restrictions, reducing daytime shadow flicker and avoiding bird and bat flying paths (see Chapter 11). Wind sector management (see Section 4.4.6) that reduces power production may be required to limit wear and tear caused by high turbulence and inflow angles, etc., in certain wind directions.

Grid regulations may also play a part in curtailments, especially where the penetration of wind power on utility systems has grown to a critical level. In Northwest China, for example, the curtailments imposed by the grid on wind power plants can be as high as 20% of the total estimated energy production. It must be questioned whether higher than 10% curtailments on the energy production would result in a profitable wind power project.

7.2   AEP Uncertainty Analysis

An uncertainty analysis is often performed as part of a wind farm energy production prediction. In concept, uncertainty is different from errors. Errors are the difference between the calculated result and the 'true value' and should be corrected if possible, whereas uncertainty is doubt about the result. We are generally more interested in the uncertainty because we typically do not know the 'true value' and an unknown error is a source of uncertainty [17]. There is always a margin of doubt in every stage of a wind resource analysis, from wind measurement to AEP predictions.

Quantifying the uncertainty is rigorously required by the industry in the planning and financing stage of wind farm projects. Understanding the common sources of uncertainty can also help to reduce their magnitude and to improve the quality of overall AEP estimates, thereby lowering the financial risk of the wind farm project. However, estimating uncertainty itself is a very uncertain business. The analyst often has to reply on his or her gut instinct to decide how confident he or she is in a particular aspect of the project. Therefore, this section is not to set anything in stone, but rather to introduce fundamentals in the estimate of uncertainty in energy yield predictions of wind farms.

7.2.1   Defining Uncertainty

We often quote uncertainty as standard deviation about the mean. Normal (also known as Gaussian) distribution is then assumed to determine the band of value for a certain confidence level. As shown in Figure 7.1, the probability density curve of a normal distribution is of a bell-like shape that is symmetric around the point $x = \mu$, defining the location of the curve. The standard deviation, σ, defines the shape of the curve, that is the spread of values. About 68% of values are likely to fall within a band of one standard deviation about the mean, typically expressed as $\mu \pm 1\sigma$. For example, if we predict that a site has an average wind speed of 9 m/s, what we actually mean is that the wind speed of the site will be somewhere under a normal bell curve centred (see Figure 7.1) at 9 m/s, and finding uncertainty really means finding the standard deviation of this normal distribution.

Broadly speaking, there are three main sources of uncertainty in AEP prediction of a wind farm. They are the uncertainty in wind measurement, prediction or wind flow modelling methodology and the natural viability of the wind parameters (e.g. wind speed and direction). The first two sources of uncertainty can be reduced through careful planning and operation by

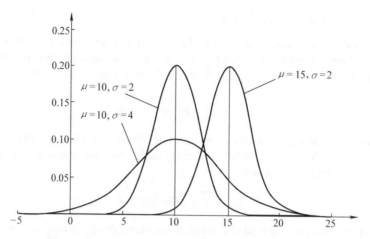

Figure 7.1   Probability density curve of a normal distribution, where μ and σ denote the mean and standard deviation respectively

limiting errors originating in equipment selection, measurement campaigns (see Chapter 8), analysis and modelling. The natural variability, however is an independent factor leading to uncertainty in long-term wind resource predictions [18].

The fitted Weibull distribution is usually used in wind resource simulation. The quality of the fit (see Section 5.5.6) should thus be considered here as an additional source of uncertainty, where a 99% coefficient of determination (R^2) roughly means a 1% $(1 - R^2)$ uncertainty in wind speed from the Weibull fit. Other sources of uncertainty involve the uncertainty in production loss factors, with, most importantly, the uncertainty in wind turbine power performance, availability and wake modelling. The uncertainty in air density can be treated as a component in power performance uncertainty, where 3% uncertainty is often assumed for standard turbine wake modelling.

In statistics, there are two types of uncertainty. Type A is the random error and uncertainty produced by variability in the quantity being measured or in the measurement procedure. The uncertainty in the estimate of the mean decreases as the number of measurements increases. The uncertainty of the mean of the quantity is equal to the standard deviation of the quantity divided by the square root of the number of repeats [19], that is

$$\sigma \bar{x} = \frac{\sigma_x}{\sqrt{N}} \tag{7.4}$$

Type B uncertainties are systematic errors or biases, often caused by an error in calibration. Effort should be made to identify the systematic errors and to either remove the source of errors or to adjust the bias. Unknown bias uncertainty is more easily characterised by an uncertainty limit, which can be converted to an equivalent standard deviation. In this case, assuming rectangular distribution, the uncertainty of the mean is given by

$$\sigma \bar{x} = \frac{\sigma_x}{\sqrt{3}} \tag{7.5}$$

7.2.2 Combining Uncertainties

Each of the independent uncertainties must be individually determined and then combined into an overall uncertainty for the whole project. Independent uncertainties are added like the sides of a triangle not like normal numbers, that is the root square sum of the components, as given by

$$\sigma = \sqrt{\sigma_1^2 + \sigma_2^2 + \cdots + \sigma_n^2} \tag{7.6}$$

For example, consider a met mast on a site for one whole year. The uncertainty of the wind measurement is 2%, but there is also an uncertainty in the variation in annual mean wind speed of 6%. The combined uncertainty is

$$\sqrt{6^2 + 2^2} = 6.3 \tag{7.7}$$

An important feature of this kind of combination is that a small uncertainty is swallowed by a large one and the overall uncertainty is much more sensitive to changes of the larger value. Figure 7.2 shows how much the combined uncertainty is reduced by taking 1% off each of the components (see Table 7.2). This example highlights the importance of performing a long-term correlation (see Chapter 6). If no correlation is performed, then the 6% uncertainty from the variation in annual mean wind speed is so large that other uncertainties are unimportant in comparison.

It should be noted that these uncertainties have to be independent in order to be added in this manner. This means that there should be no link between their values. In wind energy assessment some kind of link always exists between almost every uncertainty component, making them all dependent. However, if this dependence is minor, it is usually still reasonable to consider them as being independent [20].

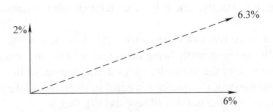

Figure 7.2 Combination of independent uncertainties

Table 7.2 Overall uncertainty reduced after taking 1% off each component

	Case I	Case II	Case III
Uncertainty in wind measurement	2%	1%	2%
Uncertainty in annual wind speed	6%	6%	5%
Combined uncertainty	6.32%	6.08%	5.39%
Overall uncertainty reduced		0.24%	0.93%

If two uncertainties are not independent and correlation can be defined between them, they should be combined through simple summation, $6\% + 2\% = 8\%$, which is very much larger. It is often a challenge as to which components are independent and which are not. Partially independent uncertainty components are usually dealt with by dividing them into separate components that are either purely independent (uncorrelated) or purely dependent (correlated).

7.2.3 From Wind Speed Uncertainty to AEP Uncertainty

Uncertainty components must be in the same unit before they are combined [17]. Most uncertainties are stated in terms of wind speed (m/s) instead of energy production (MW h) directly. The procedure of establishing the relationship is often referred to as sensitivity analysis, which converts a percentage change in the mean wind speed into an equivalent percentage change in energy production.

The relationship between the uncertainty in speed and the uncertainty in energy production varies depending on the turbine model, wind speed frequency distribution and other factors. Assume a constant Weibull k parameter. A new wind speed frequency distribution can be calculated for each wind speed bin (see Section 5.4) corresponding to the wind turbine power curve, when the mean wind speed is lowered by 1%. A new AEP is therefore computed using either Equation (7.1) or Equation (7.2) and the sensitivity, S, can be calculated as

$$S = \frac{AEP_1 - AEP_2}{AEP_1} \tag{7.8}$$

where AEP_1 is the original AEP and AEP_2 is the value after the mean wind speed has been reduced by 1%.

Then, the uncertainty in the mean wind speed can be easily converted into the uncertainty in AEP, using

$$\sigma_{AEP} = S\sigma_V \tag{7.9}$$

It should be noted that the sensitivity is not linear with respect to the mean wind speed. Therefore the crude calculation above is only valid if the wind speed does not change too much. The uncertainty module in WindPRO calculates the sensitivity for us.

7.2.4 P90, P75 and P50 AEP

Presenting project uncertainties using probabilities of exceedance in terms of expected wind farm AEP is of particular interest for risk profiling in the financial community. A 'PXX' denotes the annual energy production level that is reached with a probability of XX%. Hence, P90 means that there is a 10% chance that the P90 level of AEP will not be reached, as shown by the shaded area in Figure 7.3. The P50 is the baseline and the result of a theoretical energy yield prediction as the probability of reaching a higher or lower annual energy production is 50:50. Both P90 and P75 are widely used by banks and investors as the base for their financing decisions. They can be expressed for different periods, usually up to the wind turbine's expected lifetime [21].

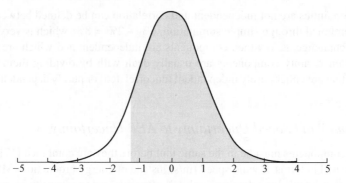

Figure 7.3   P90 (shaded area) in normal distribution

In a normal distribution, various probabilities of exceedance can be calculated by the mean and standard deviation (in the same unit as the mean) as

$$P95 = \mu - 1.96\sigma \tag{7.10}$$

$$P90 = \mu - 1.28\sigma \tag{7.11}$$

$$P75 = \mu - 0.67\sigma \tag{7.12}$$

$$P50 = \mu \tag{7.13}$$

For instance, if the base-case AEP is 136 GW h and the overall uncertainty, σ, is 12% (equivalent to 16.3 GW h, quoted as a proportion of the mean), the P95 will be 115 GW h and the P75 will be 125 GW h. With more uncertainty, those values decrease. The example curve of the expected AEP with respect to different levels of exceedance probabilities is given in Figure 7.4.

Actually, exceedance probabilities appear all over the wind industry. Most common is the use of the P90 AEP for banking projects, but the P10 is also used in the calculation of IEC turbulence. Class A turbines must be certified not to survive 16%, but $16\% + 1.28\sigma$ of turbulence intensity which is normally about 18%. This means that the turbulence will only go above this figure 10% of the time, so this is a P10 for turbulence intensity.

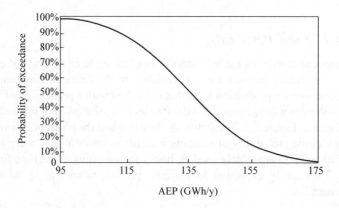

Figure 7.4   An example curve of expected AEP against probabilities of exceedance in a normal distribution

7.3   Natural Variability of Wind

The wind can fluctuate at various time scales; it is subject to diurnal and hourly changes, as well as seasonal variations of peak wind resource in winter or summer. The degree of variations is very site dependent. One year of wind measurement is usually the minimum requirement to take into account variations of intra-annual variations. Here we are interested in inter-annual variability and long-term stability of the wind.

7.3.1   Inter-Annual Wind Speed Variability

The predictions made from a one-year wind measurement may not be representative for long-term windiness. Results have shown that the inter-annual variability of the wind speed, that is the uncertainty of the annual mean wind speed, is rather site specific, ranging from 3% to 7% (usually assumed as 6%) based on reanalysis data [22], corresponding to a variation of AEP of about 8% to 18%. This represents the most significant source of variation in most cases. A correction on the short-term based prediction therefore should be made, if appropriate, in order to come to a more accurate conclusion on the average AEP for the expected lifetime of wind turbines.

The MCP techniques introduced in Chapter 6 make a long-term correction of the wind condition possible. However, correction on an AEP prediction is only recommended when the actions would lower the uncertainty level of the results. For example, if the inter-annual variability of the wind is assumed to be 6% and one year of wind data is used for the energy production estimate, then the long-term corrected wind speed should have an associated uncertainty level of less than 6%. This means that the long-term correction in general should not inflict more uncertainty on the energy production estimate unless it is verified through more than one source of long-term reference data that the on-site measurement period is obviously low or high (nonrepresentative) in windiness as compared to the long-term average.

Usually no strong underlying pattern in the annual mean wind speed can be identified, leading to the common assumption that the annual mean wind speed is statistically independent. Thus, according to Equation (7.4), the intrinsic uncertainty of the annual mean wind speed for multiple years can be computed as

$$\sigma_N = \frac{\sigma_1}{\sqrt{N_M}} \tag{7.14}$$

where σ_1 denotes the intrinsic uncertainty of the 1-year mean wind speed and N_M is the number of full years of on-site measurement.

Assuming 6% (a rather conservative assumption) uncertainty in the annual mean wind speed (1-year average), the uncertainty of the 2-year mean wind speed, for instance, is 4.24% ($6\%/\sqrt{2}$), as per Equation (7.14). Similarly, the uncertainty of the 3-year mean wind speed is 3.46% ($6\%/\sqrt{3}$). Further, assuming 80% of the correlation coefficient, R^2, for MCP and 10 years of consistent long-term reference data, then the approximate uncertainty of the long-term correction can be calculated from Equation (6.26) as 2.55% for 2-year on-site target wind data and 2.30% for 3-year on-site target wind data (assume that the entire on-site measurement period overlaps with the reference data). Therefore, the long-term correction using the MCP technique would plausibly achieve a reduction in the intrinsic uncertainty of the annual mean wind speed by about 40% and 33% respectively. The WindPRO MCP module provides a standard error on the long-term correlated wind speed, which can be converted directly to a percentage error (standard deviation) by dividing it by the mean wind speed at the measurement height.

7.3.2   Long-Term Stability of Windiness

Another point worth mentioning regarding MCP is the assumption that the wind conditions will vary in the next decades in a similar way as they did in the past during the reference period, as the past is used as a predictor for the future. However, how much will this assumption hold, given this age of global warming and climate change? Even without climate change, it is still reasonable to question the genuineness of this assumption. It is therefore important to evaluate the uncertainty associated with this assumption and to account for it in the total uncertainty of long-term corrected wind speed, that is the uncertainty of the future wind resource.

The fundamental driving force of the prevailing winds is maintained by a temperature gradient that decreases towards the poles: the stronger the temperature contrast, the stronger the wind [23]. However, the effect of global warming tends to decrease the temperature contrast slightly, as the polar regions warm up faster than lower latitudes. As that temperature contrast weakens, so do the winds. Ren [23] found that the wind power density over China, at an average height of a wind turbine, is expected to decrease by about 14% (corresponding roughly to a 5% reduction in mean wind speed) within this century. Breslow and Sailor [24] also suggested that the annual mean wind speed across the United States could decrease from 1.0% to 3.2% by 2050 and 1.4% to 4.5% by 2100, compared to the baseline period of 1948 to 1975. There are other reports supportive of the findings just mentioned. For the lifespan of wind power projects, however, which is typically 20 years, the decrease in the mean wind speed caused by global warming will be rather modest. Assuming a 0.5% (10-year project horizon) to 1.5% (20-year project horizon) uncertainty in energy production of wind parks inferred purely by global warming seems fair.

A more important factor in long-term windiness is the periodical and oscillating weather patterns, including El Niño and La Niña [25], which normally happen at irregular intervals of 2 to 7 years, averaging 5 years, and last for 9 months to 2 years, and the less famous North Atlantic oscillation (NAO) and Pacific decadal oscillation (PDO), which oscillate in long cycles of 2 to 3 decades or even longer, comparable to or exceeding the lifetime of wind power projects. El Niño refers to a band of anomalously warm ocean water temperatures that periodically develop off the western coast of South America and can cause climatic changes across the Pacific Ocean, while the coupled phenomenon, La Niña, refers to the cold temperature phase in the western Pacific. Southern oscillation (ENSO) is the oscillation of El Niño and La Niña. Ideally, the reference data period should cover at least 2 or 3 oscillation periods of ENSO, corresponding to approximately 10 to 15 years. This time horizon is practical for most MCP studies with reliable reference data. However, due to the inconsistency problem already described in Section 6.4 it is far more difficult to take into account the long-term windiness variations inferred by NAO and PDO phenomena in MCP studies. For those who are unfamiliar with climatology, NAO is the air pressure fluctuations at sea level between the Icelandic low-pressure system and the Azores high-pressure system. PDO is detected as warm or cool surface waters in the Pacific Ocean, north of 20° N, but, unlike ENSO, it is not a single physical mode of ocean variability, but rather the sum of several processes with different dynamic origins [26].

The final factor that could alter the long-term windiness in the future is the change in land cover around a wind park area. These changes can be the result of tree growing or clearing, as well as urban development. Unlike global warming and possible oscillation of weather patterns, the change in wind resource availability caused by land coverage can be speculated and even quantified in most cases, whereby corrections on the energy production estimate is possible.

The total uncertainty in long-term windiness should be an independent component representing the financial horizon (or the expected life) of the wind project. It has the same form as Equation (7.14):

$$\sigma_P = \frac{\sigma_1}{\sqrt{N_P}} \tag{7.15}$$

where N_P represents the financial horizon of the project, which typically is 20 years but may range from 10 to 25 years.

In general, 2% to 3% (corresponding to the project life of 20 years) of total uncertainty from long-term future windiness is recommended. This uncertainty is assumed to be statistically independent of the MCP uncertainty calculated from Equation (6.26) and the inter-annual uncertainty in mean wind speed defined by Equation (7.14).

7.4   Uncertainty in Wind Measurement

The wind is usually measured by 10 min averages of the wind (speed and direction) and sampled at about 1 Hz. It is then represented by a time series of these 10 min averages. The errors in the wind data measurement belong to the category of wind measurement uncertainty.

The factors that can contribute to the errors in wind measurement are anemometer uncertainty, tower shadowing effects, boom and mounting effects and data processing accuracy. For cup anemometers, the anemometer uncertainty can be further broken down to calibration uncertainty and the operational characteristics of the anemometer [18, 19]. For instance, if the anemometer is not properly mounted with sufficient separation distances between different components and optimum orientations for sensor booms, speed-up and slow-down effects can occur. There are international guidelines to follow for wind anemometry to reduce these effects. The guidelines do not eliminate these effects, but are generally designed to limit the uncertainty due to the effects to 0.5% [18]. An understanding of the mechanisms is necessary in order to minimise them in the wind measurement campaign and to identify them during wind data analysis. The mechanisms of these effects will be introduced in Chapter 8.

Other potential factors that may contribute to the quality of wind data are degradation of anemometers over time, particularly ball-bearing types and missing data due to icing or slow response to logging equipment malfunction. They all belong to the category of poor equipment maintenance. The degradation can be revealed through comparison to a backup anemometer or post-calibration of the anemometer. Long periods of missing data are costly to recover and may bring in a substantial source of uncertainty on the mean wind speed.

Table 7.3 summarises the main sources of uncertainty in wind measurement. It highlights the importance of quality control in the wind measurement campaign. If the wind data recovery rate is higher than 97% and with good correlation with wind data of other heights on the same mast, the uncertainty induced by missing data can be considered as zero; otherwise, an extra source of uncertainty should be added depending on the amount of missing data. The uncertainty of lidar/sodar measurement is not well understood and therefore is usually assumed to be higher than that of the cup anemometer.

Table 7.3   Main sources of uncertainty in wind measurement (cup anemometer)

	Standard quality	Poor quality
Calibration of the anemometer	1%	5%
Operational characteristics of the anemometer	1%	2.5%
Flow distortion due to suboptimal mounting	1.5%	3.5%
Poor maintenance of the equipment	0%	2%
Data processing accuracy	0.2%	1%
Combined	2%	7%

7.5   Uncertainty in Wind Flow Modelling

Numerical wind flow modelling is used to calculate the wind park's energy production. It is done through extrapolating the wind from the mast to every proposed wind turbine location and from the measurement height to the wind turbine hub height. The range of uncertainties induced by wind flow modelling can be very wide, as it depends on the model used, the model's resolution, the terrain and wind climate, the quality of wind measurements, the placement of the masts and many other factors. The uncertainty in wind flow modelling is a particularly important contributor to the total uncertainty, but it is often not an easy task to estimate it in a rigorous, objective fashion. Different authors and organisations may recommend quite different values and distinctive ways of defining the component, but a typical range of the combined uncertainty in wind flow modelling is 3% to 10% with respect to the wind speed, depending mainly on the complexity of the terrain and the distance to the mast [27].

It is possible to directly estimate total uncertainty in wind flow modelling if there are at least five (for statistical significance) well-distributed and representative masts within the wind turbine array of the project area with each providing sufficient data. In this case, each mast will be designated as the reference mast in turn, in order to predict the wind speeds at each of the other masts. The standard deviation of all errors between the predicted and corresponding observed wind speeds are then calculated to directly estimate the overall uncertainty in wind flow modelling [27, 28].

More often than not, there are only one or two eligible masts available in the project area, where the analyst must rely on experience and even best guesses to estimate the uncertainty. This section presents a conceptual framework for understanding wind flow modelling uncertainty. It is suggested that readers should make their own decisions based on their grasp of the project and the software tools in use, sometimes even using their gut feelings.

7.5.1   Vertical Extrapolation

Vertical extrapolation is required to estimate the wind speed at the wind turbine hub height when the wind speed is measured at a height different from the hub height. The uncertainty caused by such an extrapolation is very dependent on the predictability of the wind shear.

Wind shear depends on time (mainly due to variations in atmospheric stability), direction (mainly due to terrain effects and land cover) and wind speed. Normally, wind flow models apply only a sector-wise wind shear independent of wind speed or time of the day [18]. If the

Table 7.4 Uncertainty of the vertical extrapolation
modelled by a logarithmic profile

Terrain type	Uncertainty for every 10 m difference between the hub height and instrument height
Flat and open	0.3%
Smooth hills	0.5%
Complex	1.0%

anemometers used for the calculation of wind shear are exposed to different tower and boom effects due to different mounting setups, this should also be taken into account in calculating the wind shear uncertainty. Accuracy of instrument heights should also be verified during a site visit. Where it is not possible to verify the instrument heights, the uncertainty of wind shear can be significant. It is intuitive to say that the uncertainty in wind shear must be dependent on the complexity of the terrain and land cover as well. More complex terrain and more complex surface features should both translate into higher uncertainty in the predicted wind shear. This is especially important if the hub height is above the mast height, where an extra uncertainty component in wind shear should be considered. The magnitude of this extra uncertainty ranges from 10% to 20% of the observed wind shear depending on the complexity of the terrain and land cover [28].

Therefore, the combined uncertainty in wind shear should be in a wide range of 3% to 10%. The lower end of this range applies to the situation where the instruments are mounted properly with very similar flow distortion effects, their heights verified, flat and open terrain and a hub height not higher than the mast height.

In the next step, we need to translate the uncertainty in wind shear into the uncertainty in the predicted wind speed at hub height. The relation is given in Brower (2012) [28], which is based on the approximate exponent shear profile:

$$\sigma_{v,hh} = \left[\left(\frac{h_H}{h_2} \right)^{\Delta\alpha} - 1 \right] \times 100\% \qquad (7.16)$$

where $\sigma_{v,hh}$ denotes the uncertainty of the wind speed at hub height due to vertical extrapolation, h_h and h_2 are the hub height and the instrument height respectively and $\Delta\alpha$ is the uncertainty in the wind shear.

For example, for a hub height of 80 m and anemometer height of 50 m (with the hub height higher than the mast height), the uncertainty of the wind speed at the hub height is 1.4% to 4.8% corresponding to 3% to 10% uncertainty in the wind shear.

The exponent shear approach is straightforward, but modern wind flow modelling usually applies a logarithmic wind profile, which is more realistic and also utilises wind speed measurements of more than two heights to fit the vertical wind profile. It is also possible to verify the accuracy of a vertical wind profile by cross-interpolating the wind speeds of multiple instrument heights and adjusting the model until the predicted wind speeds match the recorded ones. The result is a smaller uncertainty in terms of vertical extrapolation of the wind speed, with the rule-of-thumb values as shown in Table 7.4.

Many analysts consider vertical extrapolation uncertainty as being separate from wind flow modelling. It is treated here as a component of wind flow modelling uncertainty as vertical extrapolation is often carried out by wind flow modelling as well. Different ways of defining the vertical extrapolation should not affect the value of total uncertainty in the AEP prediction.

7.5.2   Horizontal Extrapolation

In an ideal world, we would monitor the wind at every proposed wind turbine location, but in reality the modelling is constrained by measurements at only a few meteorological towers. Horizontally extrapolating the wind from the monitoring site(s) over the wind farm area in order to predict energy output at turbine locations is the most significant contributor to the uncertainty in wind flow modelling. It may account for one-fourth or more of the total uncertainty in AEP prediction at large and complex sites [27]. Unfortunately, however, the wind industry still lacks a rigorous analytical framework for quantifying this factor in project design and wind resource assessment, confined by the nature of numerical modelling.

The qualities of input data including a digital contour and roughness map, accuracy of turbine and mast coordinates, wind turbine characteristics, etc., are all important factors in consideration. Therefore, site visits are always recommended to verify every aspect of the project that may potentially affect the uncertainty of wind flow modelling. The optimum contour interval for a digital contour map is dependent on the nature of the terrain (complex terrain should be presented with greater detail) at the proposed wind farm and more importantly the model's resolution. The size of the simulation domain defined in the model should be sufficiently large to take into account effects from the boundaries. The assigned roughness values may vary substantially according to the observer for the same studied site, and thus should be verified through vertical interpolation between the wind speeds measured at different heights of the same mast.

Despite the lack of a rigorous analytical framework, the uncertainty component of horizontal extrapolation in wind flow modelling is usually 'guessed' according to two uncertainty components: the complexity of the terrain and distance to the mast. It is also often assumed that a wind park with a long distance from the mast to turbines on a flat terrain has a similar uncertainty to a park with a short distance and a complex terrain. We usually assign a single uncertainty value according to the average conditions of a group of turbines associated with a particular mast or for the project as a whole. In this case, we may consider Tables 7.5 and 7.6. It should be emphasised that the values given in these tables are not based on thorough studies, and they may differ substantially from case to case.

This rudimentary approach above does not consider the variation in uncertainty with position across the project area, which can lead to sizeable disparities between expected and actual production once the plants are built [27]. Therefore, it may be attractive to use Table 7.7 in order to individualise the uncertainty value for each wind turbine and then calculate the grand total. There is a fair amount of guess work regardless of which approach is to be taken. Applying Table 7.7 is a much more complicated process. The uncertainty module in WindPRO provides the possibility of using both methods.

It is generally accepted (though some may argue) that CFD-based wind flow modelling is more sophisticated in a complex terrain than WAsP and hence the reason why the uncertainties associated with CFD modelling are reduced in 7.6 and 7.7. However, one has to remember

Table 7.5 Uncertainty of horizontal extrapolation due to the average distance to the mast

Average distance to the mast	WAsP and similar	CFD models
<500 m	0.8%	0.8%
500–1000 m	1.5%	1.5%
1000–2000 m	3.0%	3.0%
>2000 m	4.5%	4.5%

Table 7.6 Uncertainty of horizontal extrapolation due to overall terrain complexity

Terrain type	WAsP and similar	CFD models
Flat and open	1.5%	1.5%
Smooth hills	2.5%	2.5%
Hills	4.0%	3.5%
Complex	6.0%	4.5%

Table 7.7 Uncertainty of horizontal extrapolation in wind flow modelling for a distance of every 1 km from the mast to the turbine in question

Terrain type	WAsP and similar	CFD models
Flat and open	0.5%	0.5%
Smooth hills	1.0%	0.9%
Complex	1.5%	1.2%

that CFD modelling is much more user-dependent than WAsP (see Chapter 3), which may introduce another unpredictable source of uncertainty.

7.5.3 Wind Resource Similarity

Uncertainty of wind flow modelling increases significantly when the topography of the wind turbine location is very different from that of the masts. In fact, the assumptions in wind flow modelling may even break down if the similarity principle is not met in some cases, resulting in a false prediction of the wind resource at wind turbine locations. A smart wind monitoring campaign design is therefore of crucial importance to lower the uncertainty in AEP prediction

Table 7.8 Uncertainty of wind flow
modelling due to wind resource similarity

Wind resource similarity	Uncertainty
Good	1.5%
Average	3.0%
Bad	5.0%

(see Section 8.1 for placement of masts). Here we may compare the met mast location with the turbine locations in respect of the terrain slope (complexity), altitude, roughness elements and obstacle influences (especially the presence of trees, the proximity to forest boundaries) and the distance to substantial roughness changes (e.g. coastlines), in order to determine the wind resource similarity. Thereafter, an uncertainty magnitude with respect to the mean wind speed may be assigned according to Table 7.8.

Most sites should be in the category of 'average'. For large, hilly sites or sites with forestry, it would often be impossible to site the met mast in a 'good' location.

7.5.4   Deploying Multiple Masts

The availability of multiple masts in the project area can considerably improve the result of the energy yield prediction while lowering the associated uncertainty, provided that the masts are representative for at least parts of the project area with sufficient wind data. They can be each assigned with a group of wind turbines in the array that are closer in distance or more similar in wind resource. The wind flow modelling is then carried out separately to realise the energy yield prediction and the corresponding uncertainty analysis for each group. The total P90 (or P70, P50) AEP will be the simple combination of the P90 (or P70, P50) AEP for each group.

A more preferable approach is to combine all the eligible masts in one modelling process and to assign weights for the masts concerning each wind turbine in the project area. This means that each turbine location will have more than one representative mast for modelling extrapolation. The weights are usually placed according to the relative distance from the mast to the wind turbine. This distance-weighted method places more weight on the mast nearer to the turbine. It assumes that the distance is the major contributor to the overall uncertainty in the AEP prediction, as the uncertainty in horizontal extrapolation is strongly related to the mast–turbine distance. As for wind resource similarity, however, the distance can be a mis-leading factor. For instance, in a coastal region, errors are likely to depend on the distance from the coast because of the influence of differences in the thermal properties and surface rough-ness of the land and the water (the internal boundary layer) instead of the distance from the mast to the turbine. Therefore, Brower et al. [27] suggested an uncertainty-weighted method that places more weight on the mast whose location is topographically most similar to the tur-bine location. Either way, there comes a question of combining the uncertainties from multiple masts according to the weighted influence of each mast on the total uncertainty.

For a single mast, it is usually reasonable to assume that different uncertainty components are independent and thus can be combined as the sum of squares. However, for different masts in the same project area, some uncertainty components may not be independent; for example natural variability of the wind is more likely to be strongly correlated across the project area. Therefore, each uncertainty component for a mast should be combined with the same component for other masts. Whether it is combined in a weighted linear sum or a weighted sum of squares depends on the definition of dependence between the masts for that individual uncertainty component.

If the uncertainty component is purely independent between masts, it can be combined using the following equation:

$$\sigma_{Combined} = \sqrt{\sum_{i=1}^{M} w_i^2 \sigma_i^2} \qquad (7.17)$$

where M denotes the number of masts associated with the wind turbine or the group of wind turbines, w_i is the weight of the mast i and σ_i is the uncertainty component of the mast i.

Assuming the same individual uncertainties, σ_0, with equal weights for all masts, Equation (7.17) is reduced to

$$\sigma_{Combined} = \frac{\sigma_0}{\sqrt{M}} \qquad (7.18)$$

For uncertainty components that are not independent, the equations should be

$$\sigma_{Combined} = \sum_{i=1}^{M} w_i \sigma_i \qquad (7.19)$$

Taking the two masts case as an example, a percentage uncertainty component is assumed to be 6% with a weight of 0.3 and 4% with a weight of 0.7 for the two masts respectively. If the uncertainty component is defined as independent between the two masts, the combined uncertainty is 3.3% according to Equation (7.17), and if it is not independent, the combined uncertainty will be 4.6% according to Equation (7.19). This example tells us that the individual uncertainty components that are independent between the masts will be reduced significantly should there be multiple proper masts available for the same project area. The combined uncertainties are to be totalled (assuming they are all independent) to come to an overall uncertainty for the whole project.

Now we have to make decisions on whether a particular uncertainty component is to be treated as independent or not between masts. It can be too optimistic to assume that all uncertainty components are independent, but doing the extreme opposite, that is assuming all errors associated with one mast are correlated with those of other masts, could result in an overly conservative evaluation of the total uncertainty as well, as it takes no credit at all for deploying multiple masts for one wind park project [28].

As a final note, attention must be paid to the possibility of systematic errors when combining uncertainties from different masts in order to avoid systematically biased results. Systematic errors may happen when, for example, the masts are situated in one type of terrain (e.g. along a ridgeline) that is not representative of some of the turbine locations. When the possibility of systematic errors is identified, the corresponding uncertainty component should not be treated as independent between masts. Multiple anemometers from the same mast are usually hardly independent in most uncertainty components, especially if they are mounted in the same direction relative to the prevailing wind, as the similar tower effects may introduce a constant bias in terms of mean wind speed and vertical extrapolation.

7.6   A Case Study

For the case study, we consider a wind farm located in an area of smooth hills, comprising 33 wind turbines with a total capacity of 49.5 MW and a hub height of 65 m. There is only one representative met mast available in the project area, providing one full year of good quality wind data. A 10-year reference wind dataset has been used in the MCP analysis that realises a correlation coefficient of 85%. The uncertainty in mean wind speed has a sensitivity of 1.6 on the AEP. The theoretical AEP computed by wind flow modelling (WAsP) is 147 GW h/year. The uncertainty component assumptions are summarised in Table 7.9. The total production loss in Table 7.9 is calculated from Equation (7.3).

Table 7.9   Uncertainty components of the case study

Uncertainty components		Production loss (%)	Uncertainty in mean wind speed (%)	Uncertainty in AEP (%)
1. Losses			**2.7**	**4.3**
1.1	Wake	−5.68	1.9	3
1.2	Availability	−5	1.3	2
1.3	Power curve performance	−3	1.3	2
1.4	Electrical	−4	0.3	0.5
1.5	Environmental	−2	0.6	1
1.6	Curtailments	0	0.0	0
2. Wind measurement			**2.0**	**3.2**
3. Wind variability			**7.0**	**11.1**
3.1	Inter-annual (1 year)		6.0	9.6
3.2	Wind in the past (MCP)		1.9	3.0
3.3	Long-term windiness		3.0	4.8
4. Wind flow modelling			**5.3**	**8.5**
4.1	Vertical extrapolation		2.0	3.2
4.2	Horizontal extrapolation: distance		3.0	4.8
4.3	Horizontal extrapolation: complexity		2.5	4.0
4.4	Wind resource similarity		3.0	4.8
5. Others			**1.3**	**2.2**
5.1	Weibull fit		0.5	0.8
5.2	Air density		1.3	2
Total		**−18.2**	**9.5**	**15.1** (18.3 GW h/y)

Table 7.10 P90 and P75 AEPs of the wind farm in the case study (GW h/y)

Project horizon	1 year	10 years	20 years
P75 AEP	108.3	110.8	110.9
P90 AEP	97.2	101.9	102.2

Table 7.11 Percentage uncertainties in AEP associated with the two masts

Uncertainties	Mast A	Mast B	Combined
Production loss	4.3	4.3	4.3
Wind measurement	3.2	3.2	2.3
Wind variability	11.1	11.1	11.1
Wind flow modelling	8.5	8.5	6.0
Others	2.2	2.2	2.2
Total	15.1	15.1	13.7

For a 10-year project lifetime, the uncertainty component 3.1 in Table 7.9 will be reduced to 1.9% in mean wind speed according to Equation (7.15). Likewise, it is 1.3% for a 20-year project lifetime. The total uncertainties for the 10-year and 20-year project lifetimes are thus 12.1% and 11.9% respectively. Without the MCP, the uncertainty component 3.2 will become 6% as only 1 year of wind measurement is available.

After deduction of the production loss, the P50 AEP becomes 120.5 GW h/y. The P90 and P75 AEP results for the wind farm project are then calculated according to Equations (7.11) and (7.12), as shown in Table 7.10.

In the case of another representative met mast available in the project area, assume that it is mounted with the exact configurations and provides wind measurements for the same one-year period as well. For the sake of simplicity for the case study, we assume 50:50 weights on the two masts for all turbines in the project. To combine uncertainties, we only assume that uncertainties in wind measurement and wind flow modelling are independent and the rest correlate between the two masts. The total uncertainty with respect to the AEP is reduced to 13.7% from 15.1% (1 year project horizon) as a result of deploying two met masts in the same project, as shown in Table 7.11.

7.7 Wind Resource Assessment Report

A wind resource assessment report provides critical information needed by wind farm developers and by banks and investors to finance a project. It summarises the entire assessment process from wind data analysis and flow modelling to uncertainty analysis. The information that should be included in the report includes the following:

1. Description of met mast(s) and wind data quality. The details of the met mast(s) are very important for the evaluation of wind data quality in terms of flow distortion induced by

the tower structure and boom. An accurate description of the mast can help us to build confidence in the wind data and explain some phenomenon in the data series.

    The coordinates of the masts should be as precise as possible and must be verified during a site visit. It is also important to specify the measurement period of the wind data and the periods with sound data recordings for a wind resource assessment.

2. Wind statistics. Once a wind data series of good quality is determined, wind statistics should be generated, usually through a software tool. The statistics tells us about the annual mean wind speed and the prevailing wind direction so that we can have a general feel about the wind resource. It is the wind statistics that is usually used in wind flow modelling as the description of the wind instead of the wind data directly.

    Other details can be useful as well, such as diurnal and annual variation of mean wind speed, correlations between two data series, wind shear, turbulence intensity, extreme wind conditions, etc.

3. Climate conditions of the wind farm. It is important to understand the natural conditions in which the turbines are to be in operation. The events like ice build-up, sandstorms in dry areas, salt spray in coastal regions and extremely high and low temperatures may have a significant impact on the practical energy yield of the wind turbines and thus should be noted carefully. Appendix III gives the list of climate conditions useful for wind resource assessment.

4. General description of the wind farm. This should include the geographical location of the wind farm, site boundaries with coordinates of turning points, altitudes, terrain type and typical roughness, land usage and prevailing weather systems, etc.

    For some sites, it may be important also to take note of potential community concerns, environmental considerations (e.g. natural reserves for rare animals)and military and aviation restriction zones that may affect the final turbine layout.

5. Wind turbine model. A few wind turbine models may be included for blind comparisons. Important information about the wind turbine models contains the power curve (which would be a site-specific annual mean air density level provided by the manufacturer), the C_t curve, the hub height and the rotor diameter. A short introduction to the power generation system can be useful for evaluating electrical loss of gross AEP.

6. Wind flow modelling. The modelling software package in use, critical parameter setups in the modelling, characteristics of the height contour and roughness maps should be noted so that readers of the report can gain a good grasp of the wind flow modelling and also so that the results can be repeated by other analyses if necessary.

7. Optimised turbine layout and the gross AEP. An optimised layout should yield maximum overall energy from the wind park while keeping the loads of every turbine under the designed limits. A wind power density map is very helpful for locating the areas of high wind resource. Great effort should be spend on lowering turbine wake losses (see Section 4.4), which are typically the dominant contributor to layout optimisation, particularly for flat terrain and offshore sites. On occasion, it can be a trade-off between energy production and turbine loads and other factors such as noise and flickering control (see Chapter 11).

8. Wind load conditions. Even though turbine loads are usually verified by the turbine manufacturer as they have to provide a product guarantee, it is reasonable to comprehend the compliance of each wind turbine with the designed limits or the IEC standard. These conditions include inflow angle, turbulence intensity at 15 m/s wind speed and exponential wind shear across the rotor diameter, particularly if the site is complex. The wind load

conditions should be sector-wise in accordance with the wind direction sectors for a wind resource assessment.

9. Uncertainty analysis. This part of the report is of particular interest for banks and investors because they will want to make sure that the value to be generated by the wind park can pay back the investment or bank load with a very high probability (P90 for instance). In other words, it makes the project a bankable one. It is unfortunately also the most arguable part of the report, simply because the uncertainty estimate itself is a very uncertain business. As a matter of fact, the entire report revolves around ensuring that the uncertainty analysis and the net AEP can be justified to be as good as possible.

10. Environmental impacts if required. Nimbyism (not in my back yard) is still the attitude of many communities on wind energy. Some of their concerns may be well justified, while others may not. It can become the major setback for a wind park project. Therefore, the report should be capable of quantifying the influences that are of particular concern for the community or the authority so that decisions can be made on actual facts. These may take account of noise and flickering impacts at certain locations and even aesthetics from certain view directions. Noise and flickering impacts are easily quantified and therefore are explained in Chapter 11. Other environmental impacts are more specific (such as impacts on birds and bats) and belong to another arena of study, and thus Chapter 11 is only able to give a fundamental introduction in order to keep the scope of the book focused.

7.8 Summary

This chapter mainly describes the calculation of a realistic energy yield of a wind park project that is acceptable for project banking and financial profiling. It starts with an evaluation of energy production losses including wake losses, availability and suboptimal power performance, etc., and the combination technique of these losses. Availability losses and losses due to suboptimal power performance of the turbines are typically the most variable ones and are usually heavily negotiated between the wind farm developer and the wind turbine manufacturer.

Once all of the losses are defined and then combined, the net AEP with 50% probability of exceedance (P50) can be realised by deducting the total loss from the theoretical gross AEP. For project financing, however, banks or investors are more conservative and would consider P50 too risky. The rest of the chapter thus focuses on building confidence on the energy yield prediction, that is an uncertainty analysis of the net AEP. The total uncertainty in wind resource assessment is broken down into many quantifiable and independent components, which can be studied separately. Generally, these uncertainty components mainly belong to three categories, that is natural viability of the wind parameters, uncertainty in wind measurements and uncertainty in wind flow modelling. By doing so, the analyst can at least have a clearer picture of where the errors and uncertainties may potentially come from and take action in every step of the wind resource assessment process, from a wind measurement campaign to wind flow modelling, in order to limit the errors or uncertainties of each component. This being so, the sense of uncertainty should become an instinct for a wind resource analyst. Anything that is not certain should be confirmed with as much precision as possible, which is why a site survey (see Section 4.4.5) becomes so important. Therefore, the tone of the book is set around reducing the uncertainty in a wind resource assessment.

The independent uncertainties are finally combined with the root-square-sum technique. In the case of deploying multiple masts, each individual uncertainty component is firstly combined between masts either in a weighted linear sum or a weighted sum of squares, depending on the definition of dependence between the masts for that individual uncertainty component.

The values of the production losses and uncertainty components are given here in a slightly conservative manner, because there is always something we do not know or we think we know in the process that is not counted in the analysis. The experiences from the past decade have also shown that we have generally overestimated the net AEP of wind park projects.

References

[1] WAsP. http://www.wasp.dk (last accessed on 30 December 2013).

[2] WindPRO. http://www.emd.dk/windpro/frontpage (last accessed on 30 December 2013).

[3] WindFarmer. http://www.gl-garradhassan.com/en/software/GHWindFarmer.php (last accessed on 30 December 2013).

[4] WindFarm. http://www.resoft.co.uk/English/index.htm (last accessed on 30 December 2013).

[5] OpenWind. http://awsopenwind.org/ (last accessed on 30 December 2013).

[6] WT. http://meteodyn.com/en/logiciels/cfd-wind-modelling-software-meteodynwt/ (last accessed on 30 December 2013).

[7] WindSim. http://www.windsim.com/ (last accessed on 30 December 2013).

[8] Kozine, I., Christensen, P., Jensen, M.W. (2000) Failure Database and Tools for Wind Turbine Availability and Reliability Analysis. Risø-R-1200(EN), Risø National Laboratory.

[9] Graves, A.M., Harman, K., Wilkinson M., Walker, R. (2008) Understanding Availability Trends of Operating Wind Farms. Poster presentation at the *AWEA WINDPOWER Conference*, Houston.

[10] IEC 61400-12-1 (2005) Wind Turbines – Part 12-1: Power Performance Measurements of Electricity Producing Wind Turbines. 2005.

[11] IEC 61400-12-2 (2013) Wind Turbines – Part 12-2: Power Performance of Electricity-Producing Wind Turbines Based on Nacelle Anemometry.

[12] Pedersen, T.F., Gjerding, S., Ingham, P., Enevoldsen, P., Hansen, K.J., Jørgensen, H.K. (2002) Wind Turbine Power Performance Verification in Complex Terrain and Wind Farms. Risø-R-1330(EN), Risø National Laboratory, Denmark.

[13] Harman, K. (2012) How Does the Real World Performance of Wind Turbines Compare with Sales Power Curves? Presented at the *EWEA 2012*, Lyon, G.L. Garrad Hassan.

[14] Hunter, R., Pedersen, T.F., Dunbabin, P., Antoniou, I., Frandsen, S., Klug, H., Albers, A., Lee, W.K. (2001) European Wind Turbine Testing Procedure Developments, Task 1: Measurement Method to Verify Wind Turbine Performance Characteristics. Risø-R-1209(EN), Risø National Laboratory, Denmark.

[15] Madariaga, A., Martínez de Ilarduya, C.J., Ceballos, S., Martínez de Alegría, I., Martín, J.L. (2012) Electrical losses in multi-MW Wind Energy Conversion Systems. *International Conference on Renewable Energies and Power Quality 2012*, Santiago de Compostela, Spain.

[16] Tamura, J. (2012) Calculation Method of Losses and Efficiency of Wind Generators, Chapter 2 of *Wind Energy Conversion Systems*, Springer-Verlag London Limited, London. DOI: 10.1007/978-1-4471-2201-2_2.

[17] Bell, S. (2001) A Beginner's Guide to Uncertainty of Measurement. National Physical Laboratory, UK.

[18] Fontaine, A., Armstrong, P. (2008) Uncertainty Analysis in Energy Yield Assessment. PB Power,K.

[19] Lackner, M.A., Rogers, A.L., Manwell, J.F. (2008) Uncertainty Analysis in MCP-Based Wind Resource Assessment and Energy Production Estimation. *Journal of Solar Energy Engineering*, **130** (3) 031006-1-031006-10.

[20] Robertson, A. (2009) *Wind and Site Uncertainty Model Manual*, Vestas Wind System A/S, Vestas Uncertainty Tool – User Manual – 20090629.

[21] Klug, K. (2006) What Does Exceedance Probabilities P90-P75-P50 Mean? *DEWI Magazine*, Nr. 28.

[22] Liléo, S., Berge, E., Undheim, O., Klinkert, R., Bredesen, R.E., Vindteknikk, K. (2013) Long-Term Correction of Wind Measurements – State-of-the-Art, Guidelines and Future Work. *Elforsk Report*, **13**, 18.

[23] Ren, D. (2010) Effects of Global Warming on Wind Energy Availability. *Journal of Renewable and Sustainable Energy*, **2** (5), 052301.

[24] Breslow, P.B., Sailor, D.J. (2002) Vulnerability of Wind Power Resources to Climate Change in the Continental United States. *Renewable Energy*, **27**, 585–598.

[25] Climate Prediction Center. Frequently Asked Questions about El Niño and La Niña. http://www.cpc.ncep. noaa.gov/products/analysis_monitoring/ensostuff/ensofaq.shtml (last accessed on 22 February 2014).

[26] Pacific Decadal Oscillation. http://en.wikipedia.org/wiki/Pacific_decadal_oscillation (last accessed on 22 February 2014).

[27] Brower, M.C., Robinson, N.M., Hale, E. (2010) *Wind Flow Modelling Uncertainty, Quantification and Application to Monitoring Strategies and Project Design*, AWS Truepower, LCC, USA.

[28] Brower, M.C. (2012) *Wind Resource Assessment, A Practical Guide to Developing a Wind Project*, John Wiley & Sons, Inc., Hoboken, New Jersey.

8

Measuring the Wind

Wind measurement campaign is the first critical step towards assessing the wind resource in a given wind park area. Measuring the wind is very expensive, not so much in terms of the cost of construction (construction costs can be significant offshore and in really complex terrain) but rather in the time it takes to complete a measurement cycle (typically a full year). After all, few organisations can afford to and are willing to waste another year on wind measurements because of bad planning and operation of a wind measurement campaign.

In an ideal world, one would measure the wind at every proposed wind turbine location and at the hub height. In reality, though, it is only possible to set up one or a few met masts in strategic location(s) to represent the wind climate of the entire wind park area. Then the wind flow modelling is supposed to take care of extrapolating the wind from the met mast location to the wind turbine locations and from the measurement height to the hub height. Therefore, one should always ask whether the met mast can represent the micro-scale wind climate of the entire project area, which is the first topic of this chapter.

The second topic of this chapter is how to ensure the quality of wind measurements. This topic has been well presented by a good number of guidelines, especially for the cup anemometer, which is still the common practice of the industry (for example see Reference [1] and [2]). Apart from the reason that it is an indispensable topic of wind resource assessment, the purpose of this chapter is to illustrate the technical aspects that could potentially impact the wind data accuracy and uncertainty, so that the wind analyst can make an educated judgement on the quality of wind measurement and limit the errors and uncertainties during the wind measurement campaign.

The content of this chapter is primarily based on cup anemometers since they have been and are still the industrial standards for wind energy applications. Alternative wind measurement techniques such as LIDAR and SODAR are also introduced in order to briefly familiarise the reader with the latest industrial trends and pros and cons of different sensors.

8.1 Representativeness of the Met Mast

The general object of an optimised wind monitoring campaign is to provide the most representative wind data covering the entire project area within budget. It can be achieved by

Wind Resource Assessment and Micro-siting, Science and Engineering, First Edition. Matthew Huaiquan Zhang.
© 2015 John Wiley & Sons, Ltd. Published 2015 by John Wiley & Sons, Ltd.

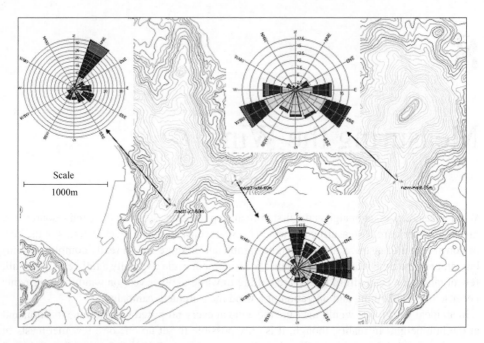

Figure 8.1   Wind condition variations on Chuandao Island in Southern China. Wind data were all measured at 70 m above ground level for a full year

placing meteorological towers (met masts) and/or ground-based remote sensing systems in appropriate locations.

Figure 8.1 shows the wind conditions measured by three met masts not far from one aother on a tropical island in Guangdong Province, China. It is very clear that wind conditions can be dramatically different within a relatively small radius, especially in the case of a complex terrain. The representative area of a single met mast may only be a few hundred metres in extreme cases.

We often use the 'similarity principle' to decide whether the met mast is representative of the turbine location. The similarity principle refers to the similarity between the met mast location and the wind turbine location in terms of the wind climate, topography and shelter effect.

8.1.1   Similar Wind Climate

Wind flow models that are applied for wind resource assessment, such as WAsP and CFD models, only simulate (meteorologicallyspeaking) microscale flow phenomena. Therefore, the inevitable hypothesis of using wind data measured at one or more locations is that the wind park area in question is within the same larger scale (meso-scale) wind climate represented by the available wind data.

8.1.1.1 Distance

Distance between the met mast and the turbine location is a very rudimentary but practical indicator of similarity in wind climates. It is recommended that the acceptable distance in a simple and uniform (Figure 2.13) terrain should be less than 10 km and less than 5 km for a simple terrain with modest height variations (Figure 2.14) and less than 2 km for a complex terrain (Figure 2.15). It is not correct to conclude that the smaller the distance the more representative is the met mast. The formation of IBL (Figure 2.2) suggests that a 10 km distance along a coastline may easily be more similar in terms of wind climate than a 3 km distance perpendicular to the coastline. A homogeneous surface property over a large area (much larger than the project area) is a prerequisite to make such a claim.

8.1.1.2 Surface Temperature

Atmospheric stability or thermal stratification of the air expresses the extent of vertical movement (convection) of the air and exerts a strong influence on the local wind climate. It is mainly determined by the temperature difference between the surface and the air immediately above it; the higher the temperature difference, the stronger is the convection and the less stable is the air.

Offshore wind monitoring campaigns are usually very expensive. Some developers may be tempted to place the met mast onshore if the offshore wind park is not far from the coast and the land area near the shore is flat and uniform. This is obviously wrong because the surface temperatures of the land and the sea are rarely the same, with each following their own variation cycles.

Large water bodies and surface areas (sand, vegetation, urban area, etc.) that may result in a difference in surface temperature should be carefully noted. It also explains from another perspective why distance alone is not able to indicate the level of similarity between the met mast location and the turbine location.

8.1.1.3 Altitude

Altitude has a profound impact on climate, most directly on air temperature and air pressure. The air flow may respond differently to the terrain at different altitudes. Vertical extrapolation of the wind climate is very difficult, especially for micro-scale wind flow modelling and should be avoided with the best endeavour.

It is generally recommended to limit the height (above sea level) difference between the met mast location and the turbine location to 100 m or 150 m maximum. In case the height variations of the project area are greater than 150 m, an extra met mast should be mounted in order to monitor the wind climate of different altitudes separately, because the chance of the project area not belonging to the same wind climate is probably too high. At the same time, the feasibility of the project area for wind park development with such great differences in elevation deserves a second thought.

The measurement height above ground level should be as close to the proposed wind turbine hub height as possible and it is suggested not to exceed 20 m between the two heights.

8.1.2 Similar Topography

The similarity in topography focuses largely on the local topographical features, primarily terrain complexity and surface roughness.

8.1.2.1 Similar Local Terrain Complexity

Terrain effects on wind flow have been introduced in Section 2.2. It intuitively implies that similar flow features (separations) should require similar levels of orographic complexity of the terrain. In general, the more complex the terrain, the smaller is the area that the met mast can represent, because the micro-scale wind climate is extremely complex in a complex terrain. Wind flow modelling tools are likely to fail to capture the real pattern of the complex wind field. With the purpose of minimizing such risks, multiple met masts are vital to represent the entire wind field.

8.1.2.2 Similar Background Roughness

Surface roughness critically defines the vertical wind profile and turbulence intensity in the surface layer. The background surface roughness can be regarded as the roughness of the dominant surface properties or roughness elements.

8.1.2.3 Similar Distance to Roughness Change Lines

An understanding of IBL (see Section 2.1.3) will naturally come to this conclusion. The most classic examples are wind parks near the shore, where the distance from the representative met mast to the coastline (roughness change line) should be similar to the distance from the proposed wind turbine position to the coastline; otherwise multiple met masts will be necessary even for a very flat terrain.

8.1.3 Similar Shelter Effect

The shelter effect from an obstacle severely alters wind flows around it. In many cases, low and prominent hills and peaks near the site can be instinctively treated as obstacles for analytical reasons (see Section 2.3). Not only does it lower the wind speed and increase turbulence intensity but it may also change the local wind direction. Figure 8.1 is a very good example.

Someone may wonder if it is a good idea to place the met mast on the medium elevation level in the terrain with large height variations in order to represent the entire area. This is not agreeable because a medium elevation level also means possible shelter impact from nearby peaks and not being able to represent the wind climate of the area at all. In other words, the surroundings of the met mast should be open and wide, particularly in the prevailing wind direction.

Sometimes it may not be possible to avoid the shelter effect from the nearby terrain, but in this case the met mast can only represent a very small area with similar shelter potentials. From a turbulence perspective, these areas are probably not feasible for placing wind turbines in the first place.

Another consideration here is to what extent the met mast can represent the wind load conditions as the shelter effect may significantly increase the turbulence intensity in the region. The assessment of wind load conditions is an essential part of the wind resource assessment as well and therefore should be represented as being possible in wind data.

8.2 Cup Anemometer Physics

A cup anemometer consists of a cup assembly (typically three) centrally connected to a vertical shaft for rotation. At least one cup always faces the oncoming wind, converting the wind pressure force to rotational torque (see Figure 8.2).

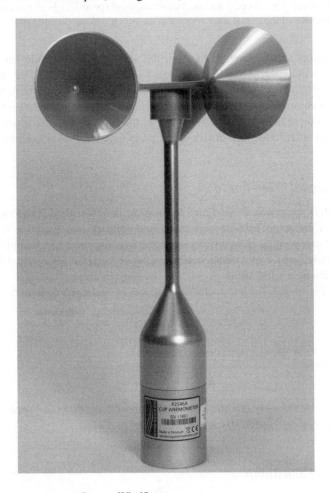

Figure 8.2 A cup anemometer *Source*: WindSenor

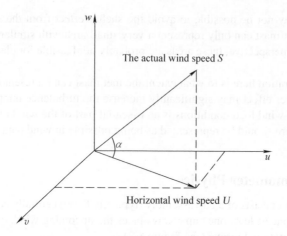

Figure 8.3 The actual wind speed S and the horizontal wind speed U, where u, v and w are the three directional components of the actual wind speed S in an orthogonal coordinate system

Cup anemometers are designed to measure the wind speed in a wide variety of applications, including wind energy, buildings, cranes, weather stations, etc. They are widely used because they are very robust, generally well suited to the definition of mean wind speed and cost attractive in comparison to other types of instrument [2], despite their generic limitations. It is therefore useful to understand the behaviour of the cup anemometer when analysing wind measurements in order to identify possible errors and quantify the uncertainty.

8.2.1 Horizontal Wind Speed

Cup anemometers are primarily designed to measure the horizontal component of the wind speed instead of the actual three-dimensional wind velocity (see Figure 8.3). They respond identically to winds coming from different directions within the horizontal plane. This is reasonable for wind energy applications because it is the horizontal wind that is available for power conversion by a wind turbine.

As shown in Figure 8.3, the horizontal wind speed, U, that cup anemometers measure is of the form

$$U = \sqrt{u^2 + v^2} \tag{8.1}$$

while the actual wind speed S is

$$S = \sqrt{u^2 + v^2 + w^2} \tag{8.2}$$

where u, v and w are the three directional components of the actual wind speed S in an orthogonal coordinate system respectively, as shown in Figure 8.3.

8.2.2 Vertical Sensitivity

The vertical sensitivity of a cup anemometer refers to its dynamic response to the vertical wind speed component. In the case of turbulence flow and flow inclination, there are always instantaneous vertical movements of the air. Papadopoulos *et al.* (2001) [3] has measured the effects of turbulence and flow inclination on the performance of a series of cup anemometers in a real flow field, showing a difference between the cups of up to 2% and errors in the measurement of the mean flow of up to 5%. The result has indicated the importance of making sure that the anemometers are mounted properly without tilting. The vertical sensitivity of cup anemometers has to do with the design of the anemometer (the rotor and the body), the wind speed and whether the anemometer is in the free atmosphere or a wind tunnel [2]. The practice guide recommended by the IEA [2] has given the test results both in the wind tunnel and in the free atmosphere for two types of anemometer in common use, as shown in Figures 8.4 to 8.7.

The IEA guide [2] further points out that cup anemometer 'A' should be used when the measurement of the full wind speed is desired and that 'B' would be preferable should the objective be to measure the horizontal plane wind. In a wind resource assessment, we usually choose the cosine vertical sensitivity profile presented by anemometer 'B' because the horizontal wind component is desired for an accurate estimate of energy production. The anemometer's sensitivity to inclined flow (off-horizontal winds) depends on the geometry of the cups.

8.2.3 *Dynamic Response in Turbulent Winds*

The term 'overspeeding' of a cup anemometer refers to the fact that it responds more quickly to positive abrupt changes in wind speed than to negative ones. This means that in a fluctuating wind, a cup anemometer is inclined to overestimate the true wind speed. Overspeeding is not considered a major source of error in the measurement of mean wind speed for IEC Class I anemometers used for turbine power performance testing and certification.

Additionally, due to the inertia of the rotor, cup anemometers cannot follow wind speed fluctuations exactly, and the higher the frequency of the fluctuation, the less able the anemometer will be to provide an accurate representation of the changes. This phenomenon is typically referred to as the 'filtering' effect. For this reason, cup anemometers tend to

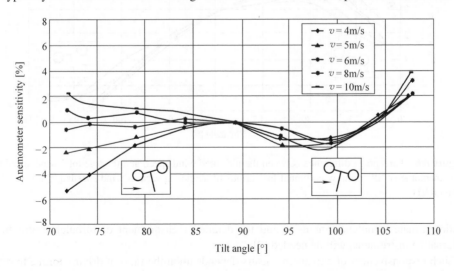

Figure 8.4 Percentage difference between the indicated wind speed and the true total wind speed for cup anemometer 'A' in the wind tunnel for various tilt angles at various wind speeds [2] *Source*: IEA Wind R&D

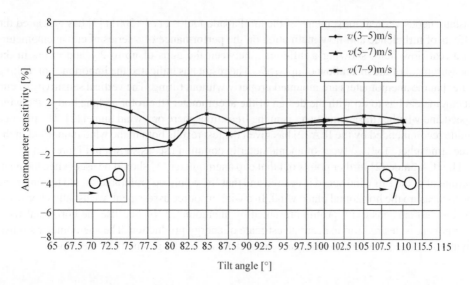

Figure 8.5 Percentage difference between indicated wind speed and true total wind speed for cup anemometer 'A' in the free atmosphere for various tilt angles at various wind speeds [2] *Source*: IEA Wind R&D

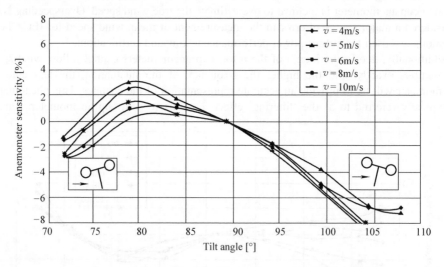

Figure 8.6 Percentage difference between the indicated wind speed and the true total wind speed for cup anemometer 'B' in the wind tunnel for various tilt angles at various wind speeds [2] *Source*: IEA Wind R&D

underestimate turbulence intensity and for detailed measurement of turbulence structure, alternative instruments will be needed [2].

Such responsiveness of cup anemometers depends upon the ratio of driving torque to rotational inertia. Driving torque increases linearly with arm length whereas inertial torque goes up to the square power. Short arms can improve responsiveness by reducing rotor inertia but will

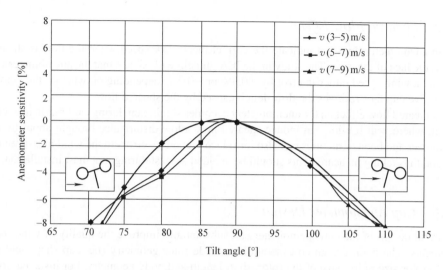

Figure 8.7 Percentage difference between indicated wind speed and true total wind speed for cup anemometer 'B' in the free atmosphere for various tilt angles at various wind speeds [2] *Source*: IEA Wind R&D

exhibit a distinct speed ripple due to flow interaction between the cups [2]. The responsiveness of the anemometer is usually indicated by the distance constant. The distance constant is the distance the air travels past the anemometer during the time it takes the cups to reach 63% of the equilibrium speed after a step change in wind speed. Longer distance constants are usually associated with heavier anemometers, which take longer to slow down when the wind speed decreases due to inertia.

Cup anemometers for use in wind energy applications typically have a distance constant of about 2.0 to 3.5 metres [2]. The anemometers with larger distance constants, such as NRG #40 and Second Wind C3, tend to overestimate the wind speed [4] in turbulent conditions than do IEC class I anemometers, depending on the degree of turbulence. It is possible to achieve a more accurate energy production estimate through applying a sensor-specific adjustment to take these tendencies into account [5, 6].

8.2.4 Nonlinearity and Mechanical Friction

Friction on the rotor shaft is always present to some degree. Friction is not linear to rotation speed. It is represented by terms factored by the speed to the zero, first and second powers. The outcome of the friction effect is that it introduces not only an offset to the calibration to overcome the zero-order term but also nonlinearity to the calibration. To make matters worse, these friction terms are temperature dependent, so an anemometer will behave differently in a warm atmosphere (e.g. summer) and in a cold atmosphere (e.g. winter). Consequently, anemometers must be calibrated with their own specific calibration certificate prior to deployment.

Friction also changes over time as the rotor shaft wears down, and changing the calibration characteristics as it does so. Post-calibration of the cup anemometer is therefore recommended so that the changes in calibration parameters can be understood; therefore the uncertainty in the wind measurements can be evaluated with more precision.

8.2.5   Sheared Flow Effect

A cup anemometer does not indicate the correct mean flow speed when the flow is sheared across the face of the cups. The study in the IEA guideline [2] shows that the anemometer will indicate a wind speed either in error by +0.7% or −0.7%, depending on whether the convex or concave cup face receives the flow deficit caused by the sheared flow.

The sheared flow effect on the anemometers becomes significant during calibration in a wind tunnel, where wall friction can induce sheared flow. Flow disturbance brought about by the mast tower and boom structures can also introduce secondary errors to the wind measurement [2], and thus taking measurements should be avoided at least during met mast installation.

8.2.6   Cup Anemometer Design

A careful design of the cup anemometer can substantially improve the quality of wind measurements. The main design considerations include rotor geometry (i.e. cup shape and the cup to rotor size ratio), size of the rotor, shaft length and body geometry. For instance, rotor responsiveness depends on the balance between aerodynamic and inertial forces, and therefore short arms (with a high cup to rotor size ratio) will give a better ability to respond to changes in the wind. Bigger and heavier rotors will have better linearity since mechanical friction will become relatively unimportant in comparison with inertia, but greater inertia will result in reduced responsiveness. Despite the absence of a formal standard for the design of

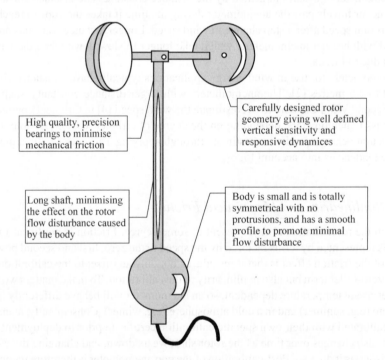

High quality, precision bearings to minimise mechanical friction

Carefully designed rotor geometry giving well defined vertical sensitivity and responsive dynamices

Long shaft, minimising the effect on the rotor flow disturbance caused by the body

Body is small and is totally symmetrical with no protrusions, and has a smooth profile to promote minimal flow disturbance

Figure 8.8   Schematic of a 'well-designed' cup anemometer [2] *Source*: IEA Wind R&D

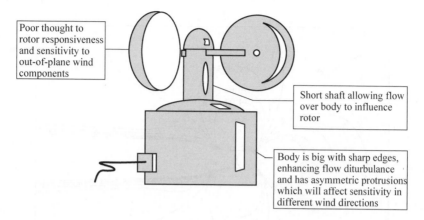

Poor thought to rotor responsiveness and sensitivity to out-of-plane wind components

Short shaft allowing flow over body to influence rotor

Body is big with sharp edges, enhancing flow diturbulance and has asymmetric protrusions which will affect sensitivity in different wind directions

Figure 8.9 Schematic of a poorly designed cup anemometer [2] *Source*: IEA Wind R&D

a cup anemometer, it is often possible to identify the general characteristics and features of a carefully designed cup anemometer [2]. The features of a well-designed and poorly designed cup anemometer are illustrated in Figures 8.8 and 8.9 respectively.

8.3   Met Mast Installation

Cup anemometers have to be deployed on a meteorological mast in order to record the wind for a long period of time. The wake of the host mast and the flow distortion induced by boom structures introduce errors in the wind measurement and the cup anemometer in this case will not be able to reflect the true free field wind speed. This being true, the separations between the anemometer and the host structures and between instruments must be adequate in order to keep such effects on an acceptably low level. The installation of a met mast has been well documented in the industry with many guidelines (such as the IEA guideline [2] and the IEC 61400-12-1 standard [7]) to refer to. These existing guidelines and standards in general aim to limit the flow distortion induced by the mast and boom structures at no greater than 0.5%. Were these standards not followed strictly, systematic errors would be introduced to the wind measurements.

8.3.1   Tower Shadow

One of the most significant sources of error in the wind measurements is typically introduced by the wake of the mast, which is usually referred to as a tower shadow. Figures 8.10 and 8.11 show iso-speed plots of the flow round a solid tubular tower and a triangular lattice tower from the IEC standards [7] (the free field wind speed comes in from the left-hand side). Both figures indicate a downward distortion of the flow upwind of the mast and a wake behind it. A certain level of acceleration can also be identified around the mast, depending on the mast structure in use.

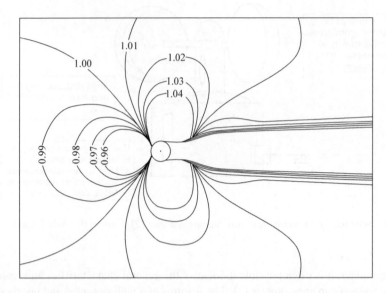

Figure 8.10   Iso-speed plot of the flow around a solid tubular tower. The local wind speed is normalised by the free-field wind speed attacking left to right [7] *Source*: IEC 61400-12-1 ed. 1.0 (2005), Annex G

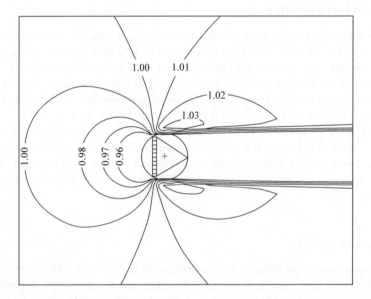

Figure 8.11   Iso-speed plot of the flow around a triangular lattice tower. The local wind speed is normalised by the free-field wind speed attacking from left to right [7] *Source*: IEC 61400-12-1 ed.1.0 (2005), Annex G

According to the IEC standards for wind turbine power performance testing,[1] the least disturbance occurs if the boom facing the wind is at 45 degrees for a solid cylinder tower (Figure 8.10), whereas in the case of the more commonly deployed triangular lattice tower (Figure 8.11), the 'free' zone is actually perpendicular to the incoming wind direction.

For a site with a relatively unidirectional wind, the boom should point to the least disturbance direction accordingly, in order to avoid the wake of the mast and flow distortions as much as possible. However, at sites where the wind is directionally spread out, the best way to avoid the wake is to have the boom pointing towards the prevailing wind direction [2]. More often than not, it is not possible to totally avoid the mast influence as the wind direction may change dramatically, particularly when the anemometer is mounted on the side instead of the top of the mast. For these reasons, the separation between the anemometer and the tower should also be adequate. A boom-mounted anemometer should thus be no closer than six mast diameters from the centre of a solid tubular mast and five mast diameters for a triangular lattice mast, provided that the boom is pointed at the least disturbance direction and the wind is very directional in nature, to achieve a less than 0.5% error induced by the tower shadow effects. The boom should be even longer if the wind is not directional and the boom has to point towards the prevailing wind direction.

8.3.2   Boom and Ancillary Effect

The direction and length of the boom is to limit the wake and flow distortion caused by the mast. However, the boom itself as well as the ancillaries (e.g. clamp and cables) may also exert surprisingly large flow distortion and should not be ignored. Things like crude and bulky clamping arrangements between the boom and the anemometer, untidy cabling, tower guys, lightening finals and other instruments may all be considerable contributors to flow distortion and thereby to measurement errors.

The degree of flow distortion cause by the boom depends on the separation of the rotor to the boom as well as the relative orientation of the boom to the wind. Experiment suggests that a rotor to boom separation of 12 to 15 boom depths are needed to keep disturbances below 0.5%, which is easily achievable in most cases [2, 8].

8.3.3   Wind Direction Vane

For cup anemometers, there has to be a separate device, that is a wind vane, for a wind direction measurement. The wind vane constantly seeks a position of force equilibrium by aligning itself into the wind (see Figure 8.12).

Wind vanes typically use a potentiometer-type transducer, which outputs a voltage signal to define the position of the vane relative to a known reference point (usually true north). There is small gap (deadband) left between the starts and ends of the reference point. The size of the

---

[1] The author thanks the International Electrotechnical Commission (IEC) for permission to reproduce information from its International Standard IEC 61400-12-1 ed.1.0 (2005). All such extracts are copyright of IEC, Geneva, Switzerland. All rights reserved. Further information on the IEC is available from www.iec.ch. IEC has no responsibility for the placement and context in which the extracts and contents are reproduced by the author; nor is IEC in any way responsible for the other content or accuracy therein. IEC 61400-12-1 ed.1.0. Copyright © 2005 IEC Geneva, Switzerland. www.iec.ch.

Figure 8.12   A wind direction vane *Source*: NRG Systems, Inc.

open deadband should not exceed 8 degrees. The resolution of the vane is recommended to be better than or equal to 1 degree. Some wind vanes divide a complete 360 degree rotation into a few equal-sized segments. This kind of resolution is too coarse for optimising the layout of a wind turbine array [4].

Because of the existence of the reference point, the alignment or the orientation of the vane becomes very important. At least the deadband area should not be aligned into or near the prevailing wind direction. The reference point is usually aligned with the boom first, so that the direction of the boom can be used to fixate the orientation of the vane. The common practice is to point the boom to the true north unless the true north is the prevailing wind direction.

It is important to distinguish the true north and the magnetic north. True north, also called the geographic north, is the direction along the local grid of longitude to the geographic north pole; the magnetic north is the direction in which the north end of a compass needle points, corresponding to the direction of the earth's magnetic field lines. It is the true north that should be considered in wind resource assessment and turbine layout design. Corrections have to be made by the angle and direction of the magnetic declination if the magnetic north is deployed during mast installation. Readers can log on to the website of the National Oceanic and Atmosphere Administration to calculate the local magnetic declination.[2]

---

[2] http://www.ngdc.noaa.gov/geomag-web/#declination.

8.3.4   Air Temperature and Other Parameters

Air temperature is used to estimate air density, which is a critical factor in wind power density. The temperature data are also important to define the operation environment of the wind farm and to detect icing events that may affect the integrity of the wind data. It is usually measured a few metres above the ground level or near the hub height. The sensor is housed within a shield to avoid being heated by direct sunlight. The sensor should be oriented towards the prevailing wind direction to ensure adequate ventilation, keeping at least one tower diameter distance from the tower to minimise tower heating effects [4]. To study atmospheric stability, two identical air temperature sensors can be mounted at two heights in exactly the same manner to measure the vertical temperature difference.

Barometric pressure can help to improve the accuracy of air density estimates slightly. Due to the dynamic pressures induced in a windy environment, high accuracy instruments are usually required. These instruments are quite expensive and, as a result, the measurement of barometric pressure is usually ignored in many wind resource assessment programmes. Instead, they usually rely either on temperature and elevation alone or pressure readings from a regional weather station for the calculation of air density.

Another meteorological parameter that is helpful for a more accurate estimate of air density is relative humidity. However, the effect of relative humidity on air density is trivial, and thus this parameter is rarely considered for the purpose of wind resource assessment.

8.3.5   Good Practice

Good practice in met mast installation can minimise the uncertainty in wind measurement. The quality of the wind measurement can also be evaluated with more precision by comparing the real installation setups with those recommended for good practice. Generally speaking, the good practice attempts to limit each uncertainty component in the wind measurement at smaller than 0.5%. The good practice is generalised in Table 8.1 and Figure 8.13. For good quality met mast installation, the combined total uncertainty in the wind speed measurement is estimated to be 1.5% to 2%, while the figure can be as high as >5% (if without proper calibration of the anemometer) in the case where the best practice has not been followed.

In Figure 8.13, the anemometer in the good practice arrangement is generally free from cables and clamps, and flow disturbance from the mast is minimised when the wind is from the prevailing wind direction. To prevent the boom from shaking at high wind speeds, a supporting triangular arm can be fixed.

Anemometers should be mounted at different height levels in order to capture the vertical profile of the wind. Redundant sensors can be deployed, particularly at the heights close to the turbine hub height, to minimise the risk of data loss due to a failed primary sensor. It is generally less expensive to deploy sensor redundancy than to replace or repair a failed sensor through an unscheduled site visit. The use of redundant wind measurement can also help reduce the uncertainty in wind resource assessment. If icing is expected, heated sensors are better used as redundant to fill the gaps in the readings when the primary sensors are influenced or even frozen by icing. Heated sensors are generally less accurate than unheated ones and consume much more power.

Table 8.1 Good practice for met mast installation

Item	Solid tubular mast	Triangular lattice mast
Anemometer boom length	>6 times mast diameter	>5 times mast diameter
Anemometer boom direction	45° to the prevailing wind direction for directional wind; pointing to the prevailing wind direction for unidirectional winds	Perpendicular to the prevailing wind direction for directional winds; pointing to the prevailing wind direction for unidirectional winds
Anemometer and boom separation	15 to 25 times the boom diameter	15 to 25 times the boom diameter
Distance between instruments	>1.5 m	>1.5 m
Vane boom direction	True north	True north

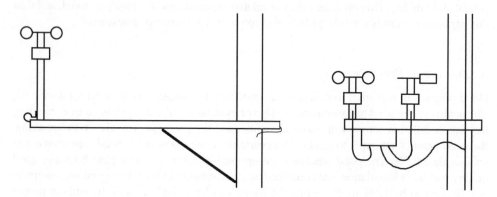

Figure 8.13   Example of good practice (left) and poor practice (right) in mounting a cup anemometer on the side of a met mast

The tower shadow effect is usually inevitable unless the anemometer is mounted on top of the mast. These effects can be easily detected in a scatter spot of speed ratios between a pair of sensors at the same height but in different boom directions. Correcting these tower effects can achieve a more accurate free-stream wind speed measurement using information provided in the IEC standard [7]. A more common practice is averaging valid data from the pair of sensors to eliminate tower effects in the combined data record. This is another reason why the deployment of redundant sensors at critical heights is so important.

Other considerations regarding met mast installation, such as lightning protection and data loggers, battery system and wirings, etc., are routine to ensure system reliability, that is the ability to constantly provide valid data without failure. These are important but do not directly relate to wind resource assessment, and therefore are not covered in detail in this chapter. Readers can refer to the manuals provided by the vendor and many guidelines available online.

8.4   Met Mast Operation and Maintenance

To ensure smooth and continuous data collection, the integrity of all system components must be well maintained and documented. Missing or erroneous data will be very expensive and time-consuming to replace. The success of the operation and maintenance programme requires simple but thorough planning and well-trained and dedicated personnel to carry out all of the procedures. To achieve this requires good documentation, scheduled and unscheduled on-site inspections to verify system integrity and a spare parts inventory in order to replace the malfunctioning instruments as quickly as possible. The essentials of such a programme are put forward in this section.

8.4.1   Documentation

Good documentation provides all the information the analyst needs to know about the met mast station and the data quality. It is essential also for an estimate of wind measurement uncertainties. The main elements that should be included in the documentation are a met mast description, data logging system and equipment list, surrounding environment and anemometer calibration reports. The following subsections give an example of the documentation elements just mentioned.

8.4.1.1   Met Mast Description

The details that should be included are listed in Table 8.2. Special attention should be paid to the accuracy of the location coordinates, which should be of a precision less than 10 m (about 0.01 min in latitude and longitude). The coordinate system in use should be indicated as well, particularly when the UTM grid coordinates are used. The dimensions of the tower and the boom can be useful for understanding possible flow disturbances that may result in less accurate readings. Individual boom directions are provided next in the equipment list.

Table 8.2 Example of a met mast description

Station name	#5 Weixian
Installation/commissioning date	13 December 2010
Location coordinates	N39 44.138 E114 46.656 (WGS 84)
Altitude (m.a.s.l.)	1940 m
Tower type	Triangular lattice
Tower height	70 m (not including lightening rod)
Tower magnitude/diameter	40 cm each side
Anemometer boom length	200 cm
Anemometer boom diameter	4 cm
Anemometer and boom separation	40 cm
Magnetic declination	6° west
Vane boom northing	True north (offset already)
Prevailing wind direction	ENE

Table 8.3   Example of a data logger system description

Logger type	NRG Symphonie
Serial number	12984
Recording interval	10 min
Wind speed calm threshold	1 m/s
Wind vane deadband	4° (true north)
Starting date	14 December 2014
Time zone	8 LST/UT

Table 8.4   Example of the equipment list

Logger channel	Senor type (unit)	Serial No.	Offset	Slope	Monitoring height (m)	Boom direction
1	Wind speed (m/s)	2144	0.769	0.436	70.4	120°
2	Wind speed (m/s)	2145	0.723	0.345	50.4	120°
3	Wind speed (m/s)	2146	0.749	0.432	30.4	120°
4	Direction (°)	3426	0	0.351	30	0
5	Temperature (°C)	5067	−86.38	0.136	3	
6	Pressure (kPa)	4066	0.043	65.09	3	
7	Voltage (V)	6036	0.021		3	

8.4.1.2   Data Logger System and Equipment List

The details shown in Tables 8.3 and 8.4 are given as an example. The time zone set up in the logger system is important for MCP analysis where the reference and the target dataset have to be aligned in the time lag. The description of the logger system should also include telecommunication information (most importantly the email address to receive data remotely, phone number to keep sufficient credit and the password) and the contact information of all stakeholders and responsible personnel at the met mast station. The offset and slope values are of particular importance for anemometers and have to individually match the results in the corresponding calibration report. If the magnetic north is used instead of true north as the reference point of boom direction, the local magnetic declination angle should be applied to the direction recordings to offset the bias.

8.4.1.3   Surrounding Environment

The description of the surrounding environment gives the analyst the first impression of the site. The best way of doing this is by taking photos in a panorama view at the site location, as shown in Figures 8.14 to 8.17.

Figure 8.14 Panorama view at the met mast location, to the north

Figure 8.15 Panorama view at the met mast location, to the east

Figure 8.16 Panorama view at the met mast location, to the south

Figure 8.17 Panorama view at the met mast location, to the west

8.4.1.4 Anemometer Calibration Reports

Calibration is critical for any measuring instrument. In the wind resource assessment, we are especially interested in the accuracy of wind speed measurements, which is why the anemometer calibration reports should be kept for data uncertainty evaluation.

Further information that can be incorporated in the documentation includes a sketch of the met mast mountings and photos of the masts and their important features.

8.4.2 On-Site Inspection

The goal of scheduled on-site inspections is to ensure the reliability of the system and detect problems promptly in order to take preventive measures. Many tasks should be carried out routinely, such as replacing batteries (or cleaning solar panels), retrieving data manually from the data logger, checking boom orientations and adjusting guy wire tensions. This is usually done once every few months, depending on the characteristics of the equipment and the manufacturer's guidelines and manuals. Unscheduled on-site inspections usually occur in emergencies to repair or replace the malfunctioning instruments as quickly as possible so that the potential loss of data can be minimised. Each on-site inspection should be well documented as well, to serve as a record of actions taken and help explain periods of questionable data. Table 8.5 lists the general activities to be taken during an on-site inspection of the met masts, some of which may require climbing the tower.

Table 8.5 Checklist of an on-site inspection

Item	Checking activity	What to do?
Location	Coordinates	Mark the coordinates using a GPS handset.
Tower	Base	Check if the bolts on the base are tight and also make sure the foundation is not sinking or distorted and free of damage.
	Straightness	Confirm the tower is straight and plumb and examine the connection points between tower sections to ensure the tower will stay straight after the inspection.
	Distortion	Confirm the tower is not twisted or distorted.
	Surface	Check for signs of rust or damage.
	Grounding system	Check if the grounding system is intact and secure and the electrical contacts are in a good condition.
Guy wires	Arrangement	The anchor points should be arranged symmetrically (120°) to avoid tension build-up and tower distortion over time.
	Alignment	The anchors of the guy wires should be well aligned to ensure that the wires in each direction are in the one vertical plane.
	Tensions	Confirm that the guy wires are properly tensioned as per the guidelines of the manufacturer and adjust the wire tension if necessary.

nce the memory card is intact and has adequate storage capacity. |

Table 8.5 (*continued*)

Item	Checking activity	What to do?
	Anchors	Check for signs of rust or corrosion and the integrity of the anchor connection, and evaluate the movement of the anchors over time.
Sensors	Boom	Check boom orientations as they may rotate and ensure that they are level and in a good condition.
	Monitoring height	Booms may slip overtime; therefore their heights should be checked to ensure that they stay the same as indicated in the documents.
	Wires and clamps	Check if the clamps are secure and if the wires and clamps could induce any additional flow disturbance to the sensors.
	Connection	Ensure that the wire connections are secure and sealed properly with silicone.
	Functionality	Check sensor outputs and make sure they are functioning properly.
	Serial number	Check if serial numbers are as reported on the documents.
Data logger	Data retrieval	Verify the email address for remote data retrieval. Ensure that the memory card is intact and has adequate storage capacity.
Data logger	Enclosure	Make sure the logger is securely locked before and after inspection. Inspect for signs of corrosion, damage, moisture and the presence of insects.
	Connection	Check the wiring panel to ensure a good connection to the sensors and that the sensors are connected to the right channel as indicated in the documents.
	Parameters	Check that the parameters (offset and scale) of the sensors are in accordance with the calibration report.
	Power	Check the battery voltage and replace them as needed. Solar panels should be cleaned and their orientation checked and examine for cracks and water resistance on the panels, wiring and electrical connection.
	Antenna	Ensure that the cellular antenna is correctly oriented

8.4.3 Monitoring

Met mast monitoring requires dedicated and well-trained eyes to promptly identify any signs of error in the data and to take measures at the soonest occasion to achieve high-quality data and low losses. They have to be observant note takers with good problem-solving abilities, responsible for documenting and explaining any periods of erroneous or lost data, as illustrated in Table 8.6. Data must be retrieved and reviewed frequently, if not daily, for obvious problems such as logger and senor failures and data transmission failures.

Table 8.6 Example of a met mast monitoring log

Time	Event	Action	Note
20 September 2010	The wind speed of 70 m height is frozen; possible damage to the sensor	Sensor replaced 27 September; the offset and scale of the new senor is 0.769 and 0.436	Typhoon occurred on 20 September and damaged the sensor
20 September 2010	50 degree discrepancy between the wind vanes of 70 m and 10 m	Boom direction of 70 m vane reoriented and secured on 27 September	Typhoon on 20 September caused the boom to rotate
10 Ocober 2010	No data received at the email address	New credit bought for the SIM card on 12 October and data transfer resumed with no data loss	RMB 200 new phone credit

8.5 Data Validation

Data validation can be a more thorough process, which ensures that the data used in wind flow modelling are valid and sound. It should be done at least quarterly to anticipate potential accumulative degradation of the sensors and detect suspect values in the data. More often, however, the analyst receives wind measurements from another party (e.g. the client or an agent) and has to validate the data comprehensively using all the documents provided (such as calibration certificates and met mast mountings) and scrutinising the data with pre-defined test criteria and in-depth graphical reviews. In this case, the analyst should pursue confirmation of key information whenever possible either through an independent party or through a site visit. Actions taken in data validation should also be documented and become a part of the wind resource assessment report.

8.5.1 Test Criteria

Test criteria can be applied in an automated programme to initially filter the data and flag suspect values. These criteria are defined by the analyst on a case-to-case basis, according to the reasonable upper and lower limits of the values (range tests), the relationship between various measured values (relational tests) and the sensible rate of change in a value over time (trend tests). Range tests are usually easier to define. For instance, the minimum value for a 10-min average mean wind speed should not fall below the anemometer offset and any values above 30 m/s should be flagged and further verified [1]; the wind direction values should be between 0 and 360 degrees. For relational tests, it is reasonable to assume, for example, that the pair of anemometers at the same height should produce similar concurrent values unless one of them is exposed to tower shadow and the wind direction values recorded by two wind vanes should have discrepancies of less than 20 degrees. Trend tests screen the values for suspect sudden changes. For example, it should be flagged when the 10 min average wind speed changes more than 5 m/s within 1 hour. It should be noted that wind direction can change abruptly during the events of frontal passage and bad weather and therefore is usually not considered as criteria [1].

As with any automated screening process, there is always a risk of overdetection (good data flagged as bad) or underdetection (bad data flagged as good), depending on the sensitivity of the defined criteria, and also the criteria for one site may not suit the other. For this reason, the analyst will have to re-examine the data commonly using wind data graphic tools. Note that verification of flagged data records may involve information outsidethe recorded data. For example, to verify if the > 30 m/s 10 min average wind speed records are valid, the analyst may need to seek extreme weather records of that day and time.

It is generally less labour intensive to scrutinise bad records for possible overdetection and spontaneously accept good records. As a result, the criteria is advised to be sensitive in order to minimise the opportunity of bad records being exempted.

8.5.2   Graphical Review

The analyst should spend a good amount of time on graphically reviewing the data records. One goal is to recognise obvious problems in the data and verify flagged data records in the automated screening process; another is to analyse error and uncertainty in the wind data and make corrections, wherever appropriate.

8.5.2.1   Time Series Graph

This is usually the first step in a data review. The entire time series of desired values is plotted with the possibility of zooming in and out for scrutiny. Problems such as icing and sensor failures can be easily detected and then excluded in the calculation. A comparison of the time series graphs of different sensors from the same mast may also alert the analyst of the need for further investigation. For instance, is the suspect trend at one anemometer (e.g. a sudden jump in wind speed) seen at other anemometers?

8.5.2.2   Scattered Plots

This is a more in-depth and a very much more flexible investigation process. The analyst may create many scatter plots to speculate on the data depending on the areas in need of examination. Wind speed ratios for a pair of anemometers at the same height can be plotted as a function of the wind direction to detect tower shadow effects and secondary tower influences. The plots of the same wind speed ratios as a function of wind speed should present a low degree of scatter of about 1; otherwise it suggests that at least one of the anemometers is not performing to its optimum. Plots of turbulence intensity as a function of the wind direction can make sense of the topography features of certain directions or suggest further investigation. These are only a few examples of what scatter plots can do.

8.5.3   Combining the Data

The target of a wind measurement campaign is to produce wind data series covering as long a period as possible with minimised uncertainty. There will be gaps left in the data series after rejecting invalid data during data validation. Filling the gaps is undoubtedly desirable for an accurate wind resource assessment, particularly for sensors at the top mast height or at the height near the turbine hub height. This is a process called data substitution.

For wind speed records, the substituted data should come from an anemometer at the same height, provided that the substituted anemometer provides valid data for the missing periods. In the case where an extended period of missing data or missing wind speed records need to be substituted with data from an anemometer from a different height, the relationship between them should be established from concurrent valid data using a simple linear regression force from the original data. At the same time, the analyst should make sure that the relationship is tight and linear in order to achieve reliable results.

If data from both anemometers at the same height are valid, it is possible to reduce the uncertainty in the observed wind speed by a factor of $\sqrt{2}$ by averaging the two datasets instead of relying on only one anemometer, assuming the measurement errors of two anemometers are independent and of approximately the same magnitude [1]. Averaging generally reduces the level of tower effects, which are only prominent in certain wind directions.

It is recommended that one of the pair of anemometers is designated as the primary one and only the invalid and missing data is replaced with valid data from the secondary anemometer, when the designated sensor has been confirmed to be of superior quality.

In terms of wind direction, the substitution is usually much more straightforward, except when two wind vanes illustrate persistent discrepancy during concurrent and valid recoding periods. Both wind vanes can be the source of bias in this case. Investigation should therefore be made to correct the bias before substituting the missing directional data.

8.5.4 Data Recovery Rate

The data recovery rate is a general indicator of wind data quality. It reflects the proportion of valid data records in the possible records over the reporting period. It is defined as

$$\text{Data recovery rate} = \frac{\text{Number of valid data records}}{\text{Number of possible data records}} \times 100\% \tag{8.3}$$

For example, if the number of possible 10-minute data records in 365 days is 52 560 but only 51 000 records are considered valid after data validation, the data recovery rate would be 97% (51 000/52 600 = 0.97).

8.6 Alternative Wind Sensors

Though cup anemometers are usually the preferred instrument, they are not without their limits, as discussed previously in this chapter. Other types of wind sensor may be preferable should the additional structure of the wind be desirable or the construction of appropriate met masts become overly expensive. In this section, a few anemometer types commonly used in wind resource assessment are introduced. Note that this section is not exhaustive.

8.6.1 Propeller Anemometer

A propeller anemometer consists of a helicoid propeller installed on a shaft that is parallel to the wind vector to be measured. The measurement of vertical wind speeds can be conducted by a vertical propeller anemometer, as shown in Figure 8.18, which should be capable of

Figure 8.18   A vertical propeller anemometer *Source*: RM Young Company

responding to both upward and downward motions. Measurement of the vertical component of the wind can be helpful for a better estimate of the energy-producing horizontal wind and characterising turbine loads.

For a propeller anemometer designed for measuring the horizontal wind component, a tail vane is mounted to keep the propeller facing to the wind direction, as shown in Figure 8.19. By using an orienting vane, the anemometer can simultaneously provide both wind speed and direction readings.

Unlike cup anemometers, overspeeding is usually not an area of concern for horizontal propeller anemometers; instead they are more inclined to underspeeding under turbulent wind conditions due to directional overshoot and misalignment from the centre of the wind direction [2]. This underspeeding effect is particularly true in low wind speeds. Stalling is an associated problem with any helicoid blades. It means that propeller anemometers cannot respond to a rapid rise in wind speed when stalling occurs. For these reasons, propeller anemometers are usually not recommended for wind turbine power performance test.

8.6.2   Sonic Anemometer

Sonic anemometers measure the wind speed and direction by detecting the variations in the speed of a high-frequency (above the range of human hearing) acoustic pulse traversed between fixed points in and opposed to the direction of the wind (see Figure 8.20). Without any mechanically rotating parts, sonic anemometers overcome many of the problems cup and propeller

Figure 8.19　A horizontal propeller anemometer with a single tail vane　*Source*: RM Young Company

Figure 8.20　A two-axis sonic anemometer　*Source*: Gill Instruments

anemometers have with respect to responsiveness to rapid wind speed and direction change. This is why sonic anemometers have been primarily deployed in the study of turbulence in the atmosphere [2].

Sonic anemometers are usually more expensive than other types and consume more power. The use of sonic anemometers is also hindered by the fact that the measurement is not well suited to the definition of mean wind speed in wind resource assessment. They may also stop working in the presence of precipitation [2].

8.6.3 Sodar

Sodar (sonic detection and ranging) is a type of ground-based remote sensing system (see Figure 8.21 for example) that measures the scattering of sound waves by small-scale fluctuations in the air density at various user-defined heights. The fluctuations in air density are caused by turbulence eddies carried along by the wind. Acoustic pulses are emitted from ground-based acoustic antennas (a series of speakers). The movement of these turbulence eddies causes a Doppler frequency shift in the return echoes (backscattering), which are received by the same antennas and then analysed by software to derive the horizontal and vertical wind velocities. When the target (a reflecting turbulent eddy) is moving towards the sodar antenna, the frequency of the backscattered return signal will be higher than the frequency of the emitted signal and vice versa [9]. The propagation time of the echoes is used to calculate the height at which scattering occurred. Sodar systems used for wind power applications are typically

Figure 8.21 A sodar system *Source*: Second Wind of Vaisala

focused on a measurement range from 50 m to 200 m above ground level, corresponding to the size of modern wind turbines.

Study has shown that the main sources of error in sodar measurements are an incorrect estimation of the zenith angle of the acoustic beams, echoes from nearby fixed objects (e.g. masts), a low signal-to-noise ratio (SNR) during times of poor temperature contrast and echoes from rain [11]. Air temperature is required to accurately calculate sound speed, as it affects both the height allocated to return echoes and the tilt angle of the acoustic beam. Precipitation can cause acoustic noise and scattering of the returned echoes, therefore invalidating the measurement of vertical wind velocity. These being true, sodar measurement during periods of precipitation are most likely to be excluded from the data analysis [1]. The deterioration of SNR for increasing height is another main problem with sodars, as it gives rise to relatively poor data quality and data recovery at higher altitudes. This issue can be addressed partly through increasing the rate of pulse repetition so that more valid data samples can be collected in each recording interval or through the pulse-coding signal processing enhancement technique, described as the next-generation bistatic sodar in Bradley (2008) [11]. Bradley has detailed the theoretical basis of the SODAR technique and its practical application.

Most sodar systems on the market nowadays are multiaxes Doppler sodar systems, meaning they are capable of detecting a signal frequency shift in three radial directions, providing a higher SNR and therefore better data quality. Compact-beam sodar systems are more accurate in a complex terrain where the wind vector can change across the measurement area of the sodar. By providing a more compact beam angle, these sodars reduce the effect of any change in the wind vector. This provides a more accurate estimate of wind flow and therefore energy production of a wind turbine. Compact beam sodars also reduce the effect of fixed echoes and allow a more compact unit design.

8.6.4   Lidar

Lidar (light detection and ranging) is another type of remote sensing system commonly used in wind energy applications. Instead of acoustic signals like sodars, lidar systems transmit a laser beam (either as a continuous wave or as a pulse) into the boundary layer and measure the Doppler shift of optical signals backscattered by aerosol particles. The technology has been in use for decades but only entered the scene of wind energy applications in recent years after sodars. Figure 8.22, for example, is a pulsed Doppler lidar designed for wind energy applications. It was first introduced in 2007 and has been tested extensively ever since.

It is intuitive that the number and size of the aerosol particles must be important for data recovery of any lidar system. As a result, the measurement of the wind at a height of over 150 m may not be possible in especially clean air. Nonuniform backscatter can occur, for example, from cloud or mist, contaminating the signals from aerosols at the height being measured. Hence a cloud correction algorithm is necessary for an accurate measurement of the wind, particularly during periods of low wind shear and high cloud cover, which is still an area of active research. The vertical downward movement of precipitation means that the measurement during periods of precipitation should be ignored, similar to the case with sodars.

Figure 8.22   A lidar system   *Source*: LEOSPHERE

8.6.5   *Deploying Sodar and Lidar*

Ground-based remote sensing, which includes sodar and lidar, has attracted increasingly more attention in recent years in wind resource assessment and turbine power performance testing. There are a few advantageous attributes that relate to the growing popularity of remote sensing, even though practical applications in wind resource assessment are still rare. First of all, as wind turbines grow larger in size and wind power projects becomes more complex, there is an inescapable need for measuring the wind at greater heights and in locations where the installation of met masts becomes prohibitively expensive or technically infeasible. Secondly, remote sensing is capable of measuring wind across the rotor area of large wind turbines and directly measuring the vertical wind profile instead of relying on wind shear extrapolation. Last, but not least, is the mobility benefit, that is they can be deployed and moved relatively easier as compared with fixed masts, which is a great advantage when it comes to verifying the wind conditions for a short period of time at a number of locations within the project area, although they do require significant more power to sustain operation than cup anemometers, especially for sodars.

On the other hand, we are still lacking in-depth knowledge of the measurement uncertainties of remote sensing and the compliance of the vector average wind speed and turbulence intensity with the IEC definitions, that is the scalar average of horizontal wind. A significant bias of

5–10% in the wind speed may occur in worst case scenarios and we need to understand where the bias comes from and how to correct it. Some of the potential errors for sodar and lidar have been discussed individually earlier in the two sections above; here we will discuss the sources of error shared by both sodar and lidar.

8.6.5.1   Remote Sensing of Wind in Complex Terrain

Unlike cup anemometers, which measure the wind at a fixed point, both sodar and lidar recover the wind measurement through averaging the wind vector over a volume of air. The depth of each averaging volume can be from less than a metre (only for lidar) to over 50 m, depending on the model in use. The averaging equations assume that each wind parcel passing over each of the scattering volumes have identical components of wind speed. This assumption of homogeneity is reasonable in a flat terrain; however, the wind can often become inhomogeneous over the spatial scale of the scattering volumes in a complex terrain. This is why in flat terrain both sodar and lidar have proven to be precise in determining mean wind speeds and wind profiles. In complex terrain, however, they have been found to have relatively large deviations (up to 10%) when compared to cup anemometers. For the same reason, in high wind shear conditions, volume averaging can result in a sizeable bias as well.

   Correction of the complex terrain influence in remote sensing is an area of ongoing research. High-resolution CFD flow modelling has been employed as a means of correcting the observed bias in a complex terrain [12, 13]. This is a much specialised area of study and thus outside the scope of this book. However, it is noteworthy that study [14] has shown that lidar errors in complex terrain correlate with RIX (see Section 2.2.4), that is low precision for high RIX.

8.6.5.2   Vector Averaging

Although the wind is a vector quantity, the wind direction and speed are treated separately as scalar values by cup anemometers. In vector averaging, like remote sensing, the orthogonal components of the wind are measured directly. The components are then summed and vector-averaged at the end of the averaging time to obtain the vector-averaged speed and direction. Because vector averaging will cancel out the winds with opposite directional components, the speeds will generally be lower than the scalar-averaged speeds. Large differences will occur with greater wind direction variance (lower wind or turbulent conditions). In terms of wind direction, vector averaging weights the direction for speed whereas the scalar-averaged direction is independent of speed, resulting in a difference of typically a few degrees [9]. Turbulence intensity measurements by remote sensing are also poorly correlated with the standard cup anemometer measurements. In the wind energy industry, it is a tradition to use scalar-averaged wind data obtained from anemometers on towers. Therefore, effort should be made to correct the vector-to-scalar bias.

   Ideally, this could be accomplished using the wind direction standard deviations measured directly from the remote sensing system. However, the measurements are generally not very accurate because they are also based on vector averages and typically use multiple beams to make the wind measurements, which do not occur at the same point in space and time. Alternatively, the standard deviations of vertical speed measured by the remote sensing system can be used to relate the vector-averaged speeds with the scalar-averaged speeds, as these measurements are generally well correlated. Further work is necessary to enable accurate measurement of the wind for wind energy purposes using remote sensing.

8.6.5.3 Interference from Obstructions

Remote sensing measures the wind in a 'cone' of air. The 'measurement cone' should be as unobstructed as possible in order to minimise harm to the data quality and maximise data recovery. It is more so for sodar systems, as these obstructions may produce noise echoes, thus reducing the SNR. Sodar systems should also stay away from high-pitched noise sources such as cars, air conditioners and generators, etc. The obstructions will become even more of a problem if they can move, like tree leaves and branches, guy wires and turbine blades. Therefore, if a reference mast is required for calibration, the remote sensing system must be placed a substantial distance from the reference mast.

8.7   Summary

This chapter explains how the wind is measured for the purpose of wind resource assessment, with the aim of helping the readers understand the uncertainties in wind measurement.

Cup anemometers have been deployed as the industrial standard in wind energy because they have been proven to be reliable for long-term wind measurement and are very economical in comparison with other types of wind sensing systems. For a successful wind measurement campaign, the resulting wind data must be representative for the wind conditions of the project area and at the same time must be accurate and reliable so that measurement uncertainties can be minimised. Cup anemometers measure the wind component in the horizontal plane instead of the true wind speed, suggesting potential errors if the cup anemometer is not placed level and in case of the presence of an off-horizontal wind component. Being mechanical means that cup anemometers are exposed to friction and inertia, which in turn give rise to nonlinear calibration characteristics and imperfect following of the changes in wind. Flow distortions from the mast and supporting structures can also introduce considerable uncertainties in the wind measurement and therefore there must be an adequate separation between the anemometer and the mast and the boom structures and so on.

Alternative wind measuring techniques are also introduced in this chapter. Remote sensing which includes sodar and lidar is explained in detail. Although remote sensing is gaining popularity, especially for sites and measuring heights that are too expensive or impractical for installing a met mast, it is not without its inherent uncertainties, which are still an area of active investigation. The parameters measured by remote sensing do not suit the standard definition of wind speed in the wind energy industry. Remote sensing measures the wind speed in a volume of air and derives the wind speed and direction using vector averaging techniques, whereas in the case of cup anemometers, the wind speed is measured and averaged as a scalar quantity. In complex terrain, inhomogeneous wind within each measured volume can cause severe errors in wind speed measurement. Correction therefore must be made on remote sensing measurements to achieve results comparable to those of cup anemometers.

References

[1] Brower, T.M.C. (2012) *Wind Resource Assessment, A Practical Guide to Developing a Wind Project*, John Wiley & Sons, Inc., Hoboken, New Jersey, USA.

[2] Expert Group Study on Recommended Practices for Wind Turbine Testing and Evaluation, 11 – Wind Speed Measurement and Use of Cup Anemometry (1999) International Energy Agency (IEA) Wind R&D, France. Available at: https://www.ieawind.org/Task_11/recommended_pract/Recommended%20Practice%2011%20Anemometry_secondPrint.pdf (last visited on 14 August 2014).

[3] Papadopoulos, K.H., Stefanatos, N.C., Paulsen, U.S., Morfiadakis, E. (2001) Effects of Turbulence and Flow Inclination on the Performance of Cup Anemometers in the Field. *Boundary Layer Meteorology*, **101**, 77–107.

[4] AWS Scientific, Inc. (1997) *Wind Resource Assessment Handbook, Fundamentals for Conducting a Successful Monitoring Program*, National Renewable Energy Laboratory, USA.

[5] Filippelli, M.V., *et al*. (2008) Adjustment of Anemometer Readings for Energy Production Estimates. *Proceedings of Windpower 2008*, Houston, Texas, USA.

[6] Hale, E., *et al*. (2010) Correction Factors for NRG #40 Anemometers Potentially Affected by Dry Friction Whip. AWS Truepower, USA.

[7] IEC 61400-12-1 (2005) Wind Turbine – Part 12-1: Power Performance Measurements of Electricity Producing Wind Turbines.

[8] Pedersen, B.M., Brinch, M., Fabian, O. (1992) Some Experimental Investigations on the Influence of the Mounting Arrangements on the Accuracy of Cup Anemometer Measurements. *Journal of Wind Engineering and Industrial Aerodynamics*, **39**, 373–383.

[9] http://www.sodar.com/about_sodar.htm, Atmospheric Research and Technology, LLC.

[10] Bradley, S., von Hünerbein, S. (2007) Comparisons of New Technologies for Wind Profile Measurements Associated with Wind Energy Applications. *Proceedings of the EWEA European Wind Energy Conference*, Milan, Italy.

[11] Bradley, S. (2008) *Atmospheric Acoustic Remote Sensing*, CRC Press, Boca Raton, FL, USA.

[12] Brady, O., Harris, M., Girault, R., Abiven, C. (2010) Correction of Remote Sensing Bias in Complex Terrain Using CFD. *Proceedings of the EWEA European Wind Energy Conference*, Warsaw, Poland, April 2010.

[13] Behrens, P. (2010) The Remote Sensing of Wind in Complex Terrain. Doctor Thesis, The Department of Physics, The University of Auckland, New Zealand.

[14] Bingol, F., Mann, J., Foussekis, D. (2009) Lidar Performance in Complex Terrain Modelled by WASP Engineering. *Proceedings of the EWEA European Wind Energy Conference*, Marseille, France, 2009.

9

Atmospheric Circulation and Wind Systems

The atmosphere surrounds us, takes up space and whose movement we feel as wind. A basic level of understanding of the physical processes and dynamical mechanisms in the atmosphere is therefore important for understanding the wind. It becomes even more important nowadays as numerical weather prediction (NWP) models enter the stage of wind resource assessment.

Atmospheric circulation and wind systems, as an important part of meteorology, have been covered by many books. This chapter will mainly present the transfer of kinetic energy in and out of the atmosphere and the actions of the driving forces and wind systems and circulations of different temporal and spatial scales, answering where the winds come from and how they are modified locally. We speak of wind systems only to facilitate their phenomenological interpretation, and one should be aware that they are not actual structures moving with the mean flow.

9.1 General Concepts

Wind and wind systems are the result of horizontal differences in air pressure created by uneven heating of the atmosphere on land and the oceans. Therefore, we can say that the sun (solar energy) is the ultimate cause of wind.

9.1.1 Vertical Structure of the Atmosphere

Though very thin compared to its planetary horizontal extent, the atmosphere has a very distinct vertical structure. The vertical structure of temperature is qualitatively similar everywhere in the atmosphere, and so it is meaningful to think of atmospheric layers through a 'typical' temperature profile (see Figure 9.1).

Wind Resource Assessment and Micro-siting, Science and Engineering, First Edition. Matthew Huaiquan Zhang.
© 2015 John Wiley & Sons, Ltd. Published 2015 by John Wiley & Sons, Ltd.

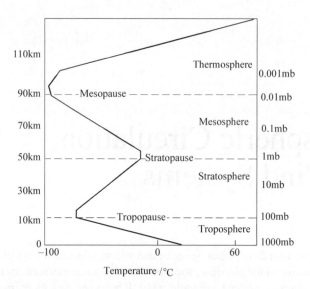

Figure 9.1   Vertical structure and temperature profile of the atmosphere

9.1.1.1   Troposphere

It is the lowest layer of the atmosphere with a depth of around 8–16 kilometres depending on the latitude and season. Temperature falls strongly moving up from the surface. This fact results from the sun's radiation striking the earth, which in turn warms the air above it. The rate of change of air temperature with height is called the 'lapse rate', which is generally about 6.5 °C per kilometre increase in altitude in the troposphere. The actual lapse rate varies with vertical mixing (warm air tends to rise and cool air tends to sink) of the air, which depends on the location, time of day, weather conditions, season, etc. The temperature can increase with height in the lower troposphere, a phenomenon called 'inversion'. 'Isothermal' is when the temperature remains the same with height. The troposphere is where weather as we know it takes place.

On the very bottom of the troposphere lies the atmospheric (also referred to as planetary) boundary layer (ABL). Its depth ranges from just a few metres to several kilometres (typically 100 to 3000 m) depending on the local meteorology. This is the part of the atmosphere that directly feels the effect of the earth's surface. It is also the part where wind energy is extracted. The wind blowing over the surface is mixed with rising air from land as it is heated by the sun, generating turbulence that redistributes heat, moisture and momentum within the ABL. The atmosphere above the ABL is often referred to as the 'free atmosphere' as it is generally free from surface forcings. The ABL will be presented in the subsequent chapter.

9.1.1.2   Stratosphere

Ranging from the top of the troposphere (marked by a temperature inversion) to 50 km above sea level, the stratosphere is a region of little mixing. This is because warmer air lying above cooler air in this region causes few overturning air currents. Transport of air and water vapour between the troposphere and stratosphere is very limited. Those particles that do manage to travel from the troposphere to the stratosphere can stay aloft for many years without returning to the ground. The rising temperature in this layer is due to the concentration of ozone absorbing ultraviolet light from the sun.

9.1.1.3 Mesosphere

The mesosphere resides from about 50 to 85 km above the earth's surface. The air pressure and density in the mesosphere are extremely low, although the percentage of oxygen is about the same, because 99.9% of the mass of the atmosphere lies below the stratopause. Without a layer of ozone, the temperature cools as the height increases.

9.1.1.4 Thermosphere

The thermosphere lies above about 90 km, marked by the percentage change of atmospheric constituents. Within this layer, ultraviolet radiation causes ionization, meaning that there is far more atomic oxygen than molecular oxygen or even nitrogen. Unlike in the troposphere and stratosphere, temperatures in the thermosphere can change by hundreds of degrees depending on the amount of solar activity.

9.1.2 Standard Atmosphere

The International Standard Atmosphere (ISA) is an atmospheric model of how the pressure, temperature, density and viscosity of the earth's atmosphere change over a wide range of altitudes or elevations. It has been established to provide a common reference for temperature and pressure and consists of tables of values at various altitudes, plus some formulas by which those values were derived. The International Organization for Standardization (ISO) publishes the ISA as an international standard, ISO 2533:1975. Other standards organizations, such as the International Civil Aviation Organization (ICAO) and the United States Government, publish extensions or subsets of the same atmospheric model under their own standards-making authority. The US Standard Atmosphere was last updated in 1976 [1].[1]

   In wind energy, the Standard Atmosphere defines the air density at standard sea level (1.225 kg/m^3) as the standard for the wind turbine power curve. In doing so, the performance of different wind turbine models can be compared under the same criteria. The definition of the standard temperature lapse rate in the troposphere (−6.5 K/km) is used to extrapolate air density from one height to another. Table 9.1 presents more details of the Standard Atmosphere.

9.1.3 Geopotential Height and Sigma Height

The two vertical coordinates are commonly used in NWP and reanalysis models. As these models gain popularity in the application of wind resource assessment, one should become familiarised with their definitions.

---

[1] 1976 Standard Atmosphere Calculator: http://www.digitaldutch.com/atmoscalc/.

Table 9.1 Standard atmosphere at sea level

Pressure P_0 (hPa)	Temperature T_0 (K)	Air density ρ (kg/m^3)	Lapse rate L (K/km)
1013.25	288.15 (15 °C)	1.2250	−6.5

9.1.3.1 Geopotential Height

Geopotential height is an adjustment to geometric height (elevation above mean sea level) using the variation of gravity with latitude and elevation. Thus it can be considered a 'gravity-adjusted height'. One usually speaks of the geopotential height of a certain pressure level (e.g. 850 hPa level), which would correspond to the geopotential height necessary to reach the given pressure. Using geopotential height eliminates air density from the primitive equations in NWP models. The equations therefore become easier to solve. It can also be used to calculate the geostrophic winds, which are faster where the contours are more closely spaced and tangential to the geopotential height contours. The geopotential thickness between pressure levels – the difference between the 850 hPa and 1000 hPa geopotential heights for example – is proportional to the mean virtual temperature in that layer.

For wind energy applications, the difference between the geopotential and geometric height is generally negligible and they can replace each other. For example, 100 m in geometric height corresponds to 99.99 m in geopotential height, a 0.002% difference; even at 5000 m, the difference is still only 0.078%.

9.1.3.2 Sigma Height

Sigma coordinates are often preferred in general circulation models because they simplify the lower boundary conditions by following the topographic variations of the earth's surface. It is defined as the ratio of the pressure at a given point in the atmosphere to the pressure on the surface of the earth underneath it. The value of sigma is between 0 and 1, with 1 representing the surface level.

For example, in WindPRO, the NCAR data are presented at the sigma level of 0.995, corresponding to 42 m above ground level, which is very close to the hub height of a typical wind turbine. Apparently, the definition of sigma height is more suitable for wind energy applications.

9.1.4 Cascade of Scales

Processes contributing to information exchange in the atmosphere have spatial scales from the size of earth down to the size of gas molecules. Limiting this to wind only, we can consider the distribution of kinetic energy in the atmosphere at various scales. This distribution is usually represented as a spectrum (see Section 9.6.3), showing how much kinetic energy there is at a certain frequency of variability (see Figure 9.2). The spectrum is continuous, which indicates that the kinetic energy cascades to neighbouring scales. For example, large planetary waves break into smaller weather systems, which in turn cause local wind circulation, which finally dissipate into small turbulent vortexes and vanish through molecular friction. Processes at

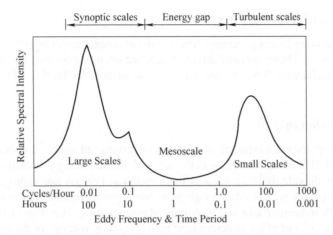

Figure 9.2 Kinetic energy spectrum near the surface in mid-latitudes [2] *Source*: Springer Science+ Business Media

different scales are driven or dominated by different forces induced by various mechanisms and therefore are studied separately.

In general, we are more interested in smaller-scale wind systems in the study of wind resource assessment. Typical weather processes of meso- and micro-scales (in the horizontal dimension) that are more or less relevant to wind resource assessment are given in Figure 9.3. According to Figure 9.3, the meso-scale involves relevant processes at between a few hundred metres and a few hundred kilometres. The lack of spectral intensity at the meso-scale, as shown in Figure 9.2, makes understanding and modelling of these processes extremely challenging.

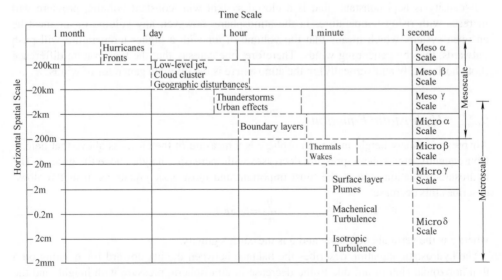

Figure 9.3 Spatial and time scales of typical meso- and micro-scale weather processes [3] *Edited from*: Orlanski (1975), reproduced with permission from Isidoro Orlanski

9.2   Laws and Driving Forces

The laws and driving forces governing winds in the atmosphere will be introduced in the following subsections. These laws and driving forces are important for understanding wind flow modelling, especially the NWP models and characteristics of the local wind resource.

9.2.1   Equation of State

Air in the atmosphere is a mixture of gasses, where nitrogen and oxygen dominate with respectively 78% and 21% of the total mass in dry air, with the rest being primarily argon (0.93%), but also carbon dioxide (0.04%), neon, helium, methane, krypton and hydrogen. For all purposes in meteorology we consider dry air as an ideal gas with molecular mass of 29 g/mol, neglecting both molecular size and intermolecular attractions. The state or thermodynamic behaviour of an amount of gas is determined by its pressure, volume (or density, which is the ratio of mass to volume) and temperature. The equation of state (also called the ideal gas law) of a hypothetical ideal gas relates these variables to one another.

There are in fact many different forms of the equation of state. The one used in meteorology is of the form

$$P = \rho R T \tag{9.1}$$

where P, T and ρ denote pressure (Pa), temperature (K) and density (kg/m^3) respectively. $R = 287.05$ J/kg K is the specific gas constant for air.

From Equation (9.1) we can conclude that when the air pressure is held constant, air density will increase with decreasing temperature. This means that air density of a site is lower in summer than it is in winter, and is also lower during the day than at night. In a vertical direction, this mechanism also facilitates convection, which is the foundation for water circulation on earth.

If density is held constant, that is a closed system with constant volume, pressure will increase with rising temperature. In the atmosphere, however, any volume of air must be an open system, which means that the volume of air with a higher temperature will push outwards, thereby generating winds. Therefore, one can say that the temperature difference (both horizontally and vertically) in the atmosphere is the ultimate generator of winds.

9.2.2   Hydrostatic Equation

Air pressure at any height in the atmosphere is a measure of the air mass above that height. Consequently, atmospheric pressure decreases with increasing height above the ground. The hydrostatic equation, one of the most important and most basic equations in meteorology, describes this decrease:

$$\frac{\partial p}{\partial z} = -g\rho \tag{9.2}$$

where z is the vertical coordinate and g is the earth's gravity.

The hydrostatic equation describes the balance between the net upward force acting on a thin horizontal slab of air, due to the decrease in atmospheric pressure with height, and the downward force due to gravitational attraction that acts on the slab, in the absence of strong vertical accelerations.

The term that varies on the right-hand side of the equation is density, which is a function of temperature (and moisture content). The air is denser in cold air, which means that in cold air the change in pressure with height is larger than in warm air, assuming the same height level. At greater heights, this decrease is smaller as air density is decreasing with height as well. The consequence of air pressure decreasing more rapidly in cold air is the horizontal pressure gradients aloft. Compensating winds would be produced to remove these pressure gradients. As the air mass aloft a warmer region moving to a colder region, the surface pressure at the warmer region would sink, which in turn leads to winds blowing from a higher to a lower pressure, that is from a colder to a warmer region on the surface. Land–sea and valley–mountain wind systems are examples of such purely pressure-driven winds.

9.2.3 Air Density

The kinetic energy content of the wind depends linearly on air density. The wind turbine power curve should also be adjusted according to site-specific air density. Following Equation (9.1), air density ρ is a direct function of air pressure p and an inverse function of air temperature T:

$$\rho = \frac{p}{RT} \tag{9.3}$$

The equation above is only valid for dry air. However, the atmosphere is rarely ever dry. Using the uncorrected temperature would give an inaccurate value for density, unless a correction was made to the gas constant for the variable contribution to density associated with water vapour. Humid air is less dense than completely dry air because water vapour has a smaller apparent molecular weight than dry air. In considering the effect of humidity, it is easier to define an artificial temperature called the virtual temperature, T_v. This is the temperature to which a dry air must be heated in order to have the same density as a humid air at the same pressure. The virtual temperature can be approximately calculated using the equation:

$$T_v = (1 + 0.608q)T \tag{9.4}$$

where q is the specific humidity (mixing ratio) of the air mass given in kg of water vapour per kg of moist air and T is in degrees Kelvin.

The virtual temperature is always slightly higher than the actual temperature, corresponding to a slight decrease in density owing to the presence of water vapour. For cold air and low specific humidity, the difference is rather small and usually neligible. However, in warm and very humid air, the increment can be sizeable. For example, for saturated humid air at 30 °C, the increment is as high as 5 K. The exact virtual temperature is also a function of air pressure.

The parameters used to calculate air density are commonly measured at a height level different from wind turbine hub heights. To vertically extrapolate air density, we also need to make use of the hydrostatic equation (9.2). From Equations (9.1) and (9.2), we get

$$\rho(z) = \frac{P_r}{RT} \exp\left(-\frac{g\,(z - z_r)}{RT}\right) \tag{9.5}$$

where P_r is the air pressure at a reference level z_r and z is the targeted height.

Using standard atmosphere at sea level as the reference (see Table 9.1), Equation (9.5) can be rewritten as

$$\rho(z) = \frac{352.9886}{T} \exp\left(-0.034163\frac{z}{T}\right) \tag{9.6}$$

The lapse rate of standard atmosphere ($L = -6.5$ K/km) can be used to calculate the temperature, T, at the targeted height:

$$T = T_r + L(z - z_r) \tag{9.7}$$

where T_r is the air temperature at a reference level z_r.

9.2.4   Forces and Winds

Air is usually considered as parcels in atmospheric physics and dynamics. A parcel is a very small amount of air, distinguishable from the rest of air, though in a continuous fluid no parcels are really independent of others. In doing so, the motion of air can be analysed from the balance of forces exerted on a parcel. According to Newton's second law of motion, an object accelerates as long as the vector sum of forces acting upon the object is different from zero. For winds in the atmosphere, the acceleration (force per unit of mass) of an air parcel caused by the balance of acting forces is

$$a = \sum F = F_p + F_f + F_r + F_c + F_g \tag{9.8}$$

where F_p is the pressure gradient force, F_f is the friction force, F_r is the Coriolis force, F_c is the centrifugal force and F_g is the gravitational force.

9.2.4.1   Pressure Gradient Force

In the atmosphere, the pressure gradient (typically of air but more generally of any fluid) is a physical quantity that describes which direction and at what rate the pressure changes the most rapidly around a particular location. The pressure gradient occurs when there is a difference in pressure between two locations in a fluid. The pressure gradient results in a net force that is directed from high to low pressure and this force is called the pressure gradient force, acting to redistribute the mass so that the pressure gradient becomes zero.

The magnitude of the pressure gradient force depends on the air density, ρ, and can be expressed as

$$F_p = -\frac{\nabla p}{\rho} \tag{9.9}$$

The minus sign is because the pressure gradient force is directed towards the lower pressure. The calculator ∇p denotes the differential pressure gradient.

9.2.4.2   Frictional Force

Friction occurs and acts as a braking force when a moving fluid is limited by a surface. For a Newtonian fluid (a good assumption for the atmosphere), the magnitude of the friction force is usually proportional to the velocity and the distance from the surface. The matter is more complicated in the atmosphere because the flow of air is turbulent and hence the notion of friction involves a description of the turbulent fluxes.

9.2.4.3 Coriolis Force

The Coriolis force (also called the Coriolis effect) is defined as the apparent deflection of objects (such as airplanes, wind, missiles and ocean currents) moving in a straight path relative to the earth's surface. It has an impact on wind, the ocean and other items flowing or flying over the earth's surface.

Its strength is proportional to the speed of the earth's rotation. As latitude increases and the speed of the earth's rotation decreases, the Coriolis force increases. The direction of the Coriolis force is perpendicular to the direction of motion. In the northern hemisphere it causes every moving object to turn towards the right, and just the opposite in the southern hemisphere. It is linearly proportional to the velocity of the moving object, v, and the Coriolis parameter, f:

$$F_r = fv \qquad (9.10)$$

For horizontal movements, the Coriolis parameter is

$$f = 2\omega \sin \varphi \qquad (9.11)$$

where $\omega = 2\pi/\text{day}$ is the angular velocity of the earth's rotation and φ is the latitude. At 45° latitude the value of f is 0.0001 s^{-1}.

For vertical movements, the Coriolis parameter is

$$f = 2\omega \cos \varphi \qquad (9.12)$$

9.2.4.4 Centrifugal Force

The centrifugal force appears when the movement of an object follows a curve instead of straight lines, pushing the object away from the centre of rotation. It is quadratically proportional to the velocity and inversely proportional to the radius of the trajectory, r, as given by

$$F_c = \frac{v^2}{r} \qquad (9.13)$$

It is important to mention that among all the four forces acting on a parcel of air, the pressure gradient force is the only one that does not require motion to exist. As a result, this is the only force in the atmosphere actually initiating motion. All the other three act to change the path of the motion of the air after the motion has been initiated by the pressure gradient force. Again, this makes the temperature difference the ultimate driver of winds in the atmosphere.

9.2.4.5 Gravitational Force

This is related to the mass of the air and is directed downwards perpendicular to the earth's surface, preventing the atmosphere from escaping into space. This is usually counterbalanced largely by the upwards pressure gradient forces and is only accountable in the case of significant vertical accelerations, for example in convective clouds.

9.3 General Atmospheric Circulations

From anunderstanding of the balance of forces acting on an air parcel, we are now able to explain large-scale winds in the atmosphere.

9.3.1 Geostrophic Winds

Geostrophic winds are the easiest and the most fundamental balance of forces found in the atmosphere. It is the theoretical wind that would result from an exact balance between the Coriolis force and the pressure gradient force on horizontal surfaces, that is both the frictional force and the centrifugal force are negligible and the gravitational force is ignored. In the equilibrium state, we have

$$F_p + F_c = 0 \qquad (9.14)$$

This condition is called the geostrophic balance, which can be found in the free atmosphere above the atmospheric boundary layer, as the frictional force from the surface disappears there and the motion of the wind can be considered straight (away from pressure maxima and minima). Much of the atmosphere outside the tropics is close to geostrophic flow much of the time.

The process of reaching the balance of these two forces is illustrated in Figure 9.4. Due to the existence of the horizontal pressure gradient (on a constant height surface), the parcel starts to move from high to low pressure, gaining speed. The Coriolis force perpendicular to the wind direction begins to act because of the motion, changing the path of the wind while increasing in magnitude with accelerating wind speed. At a certain point, the wind blows in a straight line parallel to the isobars with zero acceleration because the pressure gradient force and the Coriolis force are balanced, that is geostrophic wind is formed.

Taking Equations (9.9) and (9.10) into (9.14) gives

$$-\frac{\nabla p}{\rho} + fv = 0 \qquad (9.15)$$

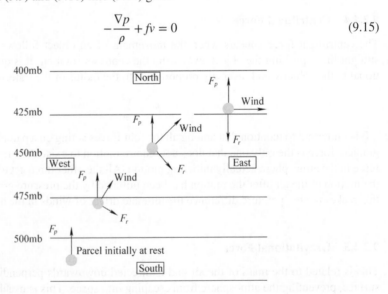

Figure 9.4 The process of pressure gradient force and Coriolis force reaching equilibrium on a constant height surface in free atmosphere aloft in the northern hemisphere

Thus

$$v = \frac{\nabla p}{\rho f} \qquad (9.16)$$

Therefore, the velocity of geostrophic winds depends on the magnitude of the large-scale horizontal pressure gradient and the air density, as well as the latitude-dependent Coriolis parameter.

Following Equation (9.16), a pressure gradient of about 1 hPa per 100 km in mid-latitudes leads to a geostrophic wind speed of about 10 m/s. Geostrophic winds exist in locations where there are no frictional forces and the isobars are straight. However, isobars are almost always curved and are very rarely evenly spaced. In the northern hemisphere, the geostrophic wind blows counterclockwise around low-pressure systems parallel to the isobars. The rotation is opposite in the southern hemisphere.

In the case of considerably curved isobars, the centrifugal force is no longer negligible. The equilibrium wind is the so-called gradient wind. It is slightly lower around low-pressure systems (because the pressure gradient force is opposite to the centrifugal force) and higher around high-pressure systems (because the pressure gradient force and the centrifugal force act in the same direction) than the geostrophic wind [4].

In rare cases, the curvature of the isobars can be so strong that the centrifugal force is much larger than the Coriolis force and the equilibrium is governed by the pressure gradient force and centrifugal forces only. Also at the equator, Equation (9.16) possesses singularity because the Coriolis parameter is zero, so the pressure gradient force must replace the Coriolis force to balance the pressure gradient force. This kind of wind, called cyclotrophic wind, is found in tornadoes and tropical cyclones near the equator.

9.3.2   Baroclinic Atmosphere and Thermal Winds

The geostrophic wind is an idealised wind assuming barotropic atmosphere within which the geostrophic wind is independent of height and the slope of isobaric surfaces is independent of temperature. This does not necessarily hold true in reality, as horizontal temperature gradients cause the thickness of gas layers between isobaric surfaces to increase with higher temperatures. As a result, the slope of isobaric surfaces increases with height (see Figure 9.5), that is the horizontal pressure gradient becomes height-dependent. This causes the magnitude of the geostrophic wind to increase with height (also see Figure 9.5). Such an atmosphere is called baroclinic.

Thermal wind is a vertical shear in the geostrophic wind caused by a large-scale horizontal temperature gradient. It is not a wind per se, but rather a wind shear. Thermal wind is a general phenomenon because the real atmosphere is nearly always slightly baroclinic [4]. If a component of the geostrophic wind is parallel to the temperature gradient, the thermal wind will cause the geostrophic wind to rotate with height. If the geostrophic wind blows from cold air to warm air (cold advection) the geostrophic wind will turn counterclockwise with height, a phenomenon known as wind backing. Otherwise, if the geostrophic wind blows from warm air to cold air (warm advection) the wind will turn clockwise with height, also known as wind veering. Wind backing and veering allow us to estimate the horizontal temperature gradient with data from an atmospheric sounding.

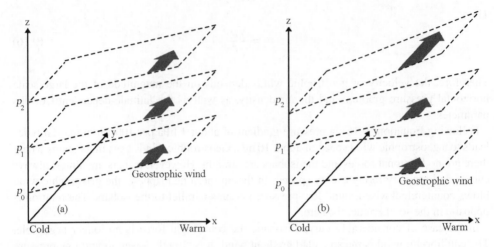

Figure 9.5 The vertical variation of geostrophic wind in a barotropic atmosphere (a) and in a baroclinic atmosphere (b). The dotted lines enclose isobaric surfaces, which remain at a constant slope with increasing height in (a) and increase in slope with height in (b)

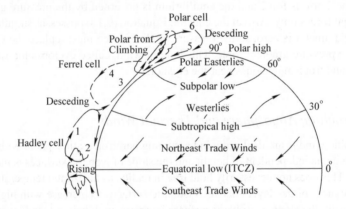

Figure 9.6 Three cell circulation in the northern hemisphere

9.3.3 Three Cell Circulation

Due to the small inclination of the earth axis of rotation, most of the solar energy hits the earth in the tropical belt, between ±30 degrees of latitude. Atmospheric circulation is the large-scale movement of air acting like a heat engine redistributing this energy on the surface of the earth. If the earth were not rotating, the circulation would be straightforward, with warm air rising in low latitudes and cold air sinking at high latitudes. However, this is not quite what happens. Instead, planetary rotation would cause the development of three circulation cells in each hemisphere rather than one (see Figure 9.6). These three circulation cells are known as the Hadley cell, the Ferrel cell and the polar cell.

The circulation cell closest to the equator is called the Hadley cell (from the equator to about 30° latitudes). The air in contact with the surface is heated by conduction in the tropics, causing a rising motion called convection. Convection is further intensified by condensation, which releases large amounts of latent heat, that is the energy released due to changes from water vapour to liquid water droplets, causing tropical thunderstorms to develop. Tropical convection removes huge quantities of heat from the surface all the way up to the top of the troposphere some 14 km high, essentially providing the energy to drive the Hadley cell. The tropical convection belt, the so-called inter-tropical convergence zone (ITCZ), draws in surface air from the subtropics, as the air rises into the upper atmosphere because of convection, producing equator directed flow near the surface. The flow near the surface is deflected westwards by the Coriolis force to create the Northeast Trades (right deflection in the northern hemisphere) and Southeast Trades (left deflection in the southern hemisphere).

At the cloud top level, however, the air has to flow away from the equator. The Coriolis force also causes the deflection of this moving air in the upper atmosphere, and by about 30° of latitude the air begins to flow zonally from west to east and no longer flows meridionally. Some of the air in the upper atmosphere has to sink back to the surface to compensate for the accumulation of air due to the zonal flow, also known as the subtropical jet stream, creating the subtropical high-pressure zone. Clear skies generally prevail throughout the surface high pressure, which is where many of the deserts are located in the world. On the surface, a portion of the sinking air diverges from this zone of high pressure towards the equator, completing the Hadley cell, while the rest of the air diverges towards the poles. Once again, the poleward flow is deflected by the Coriolis force, producing Westerlies roughly from 30° to 60° latitude, taken over by a new cell known as the Ferrel cell. At 60° North and South latitude, the Westerlies are still warmer than the air approaching from the poles, resulting in frontal uplift and cloud formation. Some of the lifted warm air returns to 30° latitude to complete the Ferrel cell. However, most of it is directed to the polar vortex, where it sinks again to create the polar high. Above 60° latitude, the polar cell circulates cold polar air towards the equator. Taking into account the effects of the Coriolis force, the winds above 60° latitude are prevailing Easterlies.

The two air masses at 60° latitude do not mix well and form the polar front, separating the warm air from the cold air. Thus the polar front[2] is the boundary between warm tropical air masses and the colder polar air moving from the north. Because the earth is rotating, the boundary between the two air masses of different density remain sloped, resulting in horizontal pressure gradients that drive the general westerly winds.

The polar jet stream aloft is located above the polar front and flows generally from west to east. It is strongest in the winter because of the greater temperature contrasts than during the summer, reaching speeds of up to 100 m/s. Because of the strong temperature gradient, this flow is unstable for initial perturbations, that is small perturbations quickly grow into large horizontal waves, which eventually break. This is known as baroclinic instability. The polar front therefore can be thought of as the average expression of the transient frontal systems that move along with mid-latitude cyclones, exhibiting a rather fragmented and wavy pattern (see Figure 9.7). The growth of perturbations due to baroclinic instability together with Hadley cells dominantly drive the meridional distribution of energy on earth (ocean currents also contribute).

---

[2] The use of the word 'front' is from military terminology; it is where opposing armies clash in battle.

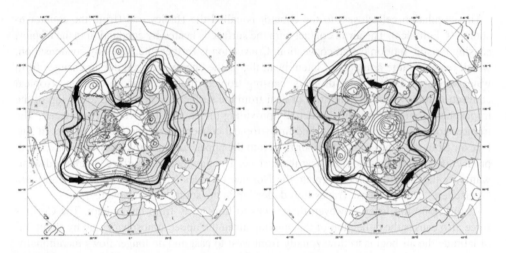

Figure 9.7 Polar front snapshots of 26 May 2006 (left) and 30 May 2006 (right) in the northern hemisphere

As a last note, the Hadley, Ferrel and polar cells running along north–south lines are major factors in global heat transport; they do not act alone. For example, the Walker circulation is an east–west circulation. It comes about because water (the ocean) has a higher specific heat capacity than land and thereby absorbs and releases more heat, but the temperature changes less than land.

9.4 Synoptic Scale Wind Systems

Synoptic scale is a term applied to large scale weather systems that span from several hundred to several thousand kilometres, corresponding to the scale typical of high- and low-pressure systems and mid-latitude depressions. Each time you watch your local weather forecast, you are seeing synoptic scale meteorology. Synoptic weather systems associated with baroclinic instability are of great importance for wind energy because the wind would otherwise be limited to the narrow region of the polar front. Most of the kinetic energy in the atmosphere is generated by wind systems on this scale. We use the term 'wind system' here rather than 'weather system' because we are looking at the weather from the perspective of winds.

Cyclones, anticyclones and weather fronts to be introduced in the first two subsections are the general mechanisms responsible for conversion of atmospheric potential energy into kinetic energy in mid-latitudes, while in tropical areas such energy conversion is dominated by tropical stroms.

9.4.1 Mid-latitude Cyclones and Anticyclones

Without surface friction, the winds would move counterclockwise around the centre of the low-pressure systems in the northern hemisphere due to the balance of pressure gradient forces and Coriolis forces. However, the existence of surface friction slows down the winds, causing the Coriolis force to weaken and the pressure gradient force to dominate. The air therefore spirals into the centre of a low-pressure system at the surface. This surface convergence leads to large-scale upward motion of the air, creating clouds and even causing rain and storms to form. Such low-pressure systems are called cyclones. Mid-latitude synoptic scale cyclones typically measure around 1000–2000 km across. Smaller cyclones also occur, but are generated by the meso-scale mechanisms and are called meso-cyclones.

On the contrary, the anticyclones produce wind circulating around high-pressure systems in a clockwise direction in the northern hemisphere. The weakened Coriolis force due to frictional effects results in the surface wind spiralling away from the centre. The air at the surface is then replaced by sinking air from above to subsidise the divergence, supressing cloud development and giving clear skies. The downward subsiding air is adiabatically compressed and heated by the resulting increase in atmospheric pressure, and as a result the lapse rate of temperature is reduced. If the air mass sinks low enough, the air at higher altitudes becomes warmer than at lower altitudes, producing a temperature inversion called subsidence inversion in meteorology. Subsiding air almost never continues downward to the earth's surface because it is almost always conteracted by turbulent mixing, however slight, near the surface [5]. Thus, a pronounced temeperature inversion above the boundary layer is a common feature in an anticyclone. A subsidence inversion differs from a surface temperature inversion in that the later one is typically due to cooling of the air near the surface due to radiative heat loss at night.

The maximum pressure gradient in a cyclone (surface convergence) is usually much stronger than in an anticyclone (surface divergence), and so are the winds. It is not uncommon even in mid-latitude cyclones for winds to sustain a hurricane force for a longer time. Flow structures of cyclones and anticyclones are illustrated in Figure 9.8.

9.4.2   Weather Fronts

A weather front is a boundary separating two masses of air of different densities, and is the principal cause of meteorological phenomena. The frontal zone is defined as a region of sharp changes in relevant atmospheric variables such as temperature, wind speed, direction and

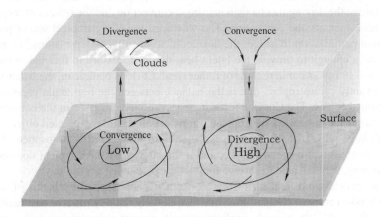

Figure 9.8   Flow structures of a cyclone (left) and an anticyclone (right)

humidity. It is represented by the leading edge of a wedge of cold/cool air. If the wedge is moving into an area previously occupied by warmer air, the front is called a cold front. A warm front occurs when the opposite movement of the wedge happens. An occluded front is when a cold front overtakes a warm front in a cyclone. A front that does not move or barely moves is called a stationary front, usually featured by winds of different velocity on each side blowing parallel to the front, which is why it is also known as a shearline.

In cold fronts, the heavier cold air directly forces warmer and often humid air to rise. Updrafts along the cold front intensify the general rising of the air in the cyclone. This may produce narrow bands (around 10–100 km) of thunderstorms and severe weather. It is therefore very common at the passage of a cold front to experience increasing wind speeds, sometimes extreme winds, a change in wind direction and a temperature drop. Though thunderstorms pose a great danger for wind turbines, the place and time of their occurrence is difficult to predict. They are thus usually dealt with in a probabilistic manner, as there is a likelihood of a thunderstorm occurring over a certain area during a certain time in the near future.

Cold fronts and occluded fronts generally move from west to east, while warm fronts move polewards. A warm front moves more slowly than a cold front because cold air is denser and harder to remove from the earth's surface. As a result, the cross-section of the active part of the warm front is usually broader than that of the cold front. Also, in general, thunderstorms do not develop along a warm front, winds change less dramatically and precipitations are less intense. Warm fronts are usually preceded by stratiform precipitation and fog. Clearing and warming is usually rapid after a frontal passage. If the warm air mass is unstable, thunderstorms may be embedded among the stratiform clouds ahead of the front and after a frontal passage thundershowers may continue [6].

9.4.3 Tropical Storms

A tropical storm is referred to by various names such as a hurricane, typhoon, tropical cyclone or tropical depression. Most tropical storms form near the equator (ITCZ) or the monsoon trough. Due to the lack of large and persistent temperature gradients in the tropics, paramount examples of tropical wind systems on a large scale, the tropical storms, require another growth mechanism, which is different from that for mid-latitude cyclones and fronts. In the case of tropical storms, the driving mechanism is instead convection and the release of large amounts of latent heat intensify the convection. The westward travelling tropical waves provide sufficient initial perturbation to trigger the development of tropical depressions.The sea surface must be warm enough to provide the latent heat that is necessary to fuel the tropical storm. Since tropical storms occur in regions of rather small Coriolis parameter, the balance of forces is predominantly cyclostrophic, that is the balance between the horizontal pressure gradient and centrifugal forces. The Coriolis force increases moving away from the equator and consequently wind speeds decrease. Tropical storms act almost like separated objects travelling along with large-scale mean flows, that is the streams in the earth's atmosphere. With sufficient fuel, that is a warm ocean surface, and if the large scale environment does not act in a destructive way, they are able to survive long journeys that last several days.

The near-surface wind field of a tropical storm is characterised by bands of air rotating rapidly around a centre of circulation, while also flowing radially inwards to a region of calm weather in the centre (the eye). As air flows radially inward, it begins to rotate counterclockwise in the northern hemisphere and clockwise in the southern hemisphere. At an inner radius,

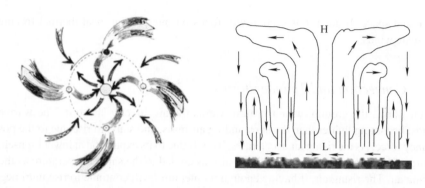

Figure 9.9 The structure of a typical tropical storm

which typically is coincident with the inner radius of the eyewall and where maximum winds are found, air begins to ascend to the top of the troposphere. Once aloft, air flows away from the storm's centre, producing a shield of cirrus clouds [7] (see Figure 9.9).

In general, wind speeds are low at the centre, increase rapidly moving outwards to a radius of maximum winds and then decay more gradually with radius to larger radii. Because of the cyclonic rotation, the wind in the tropical storms is stronger on the right-hand side looking in the direction of storm propagation. Additional spatial and temporal variability may occur due to the effects of localized processes, such as thunderstorm activity and horizontal flow instabilities. In the vertical direction, winds are strongest near the surface and decay with height within the troposphere [7]. Due to vicious convection and high humidity, tropical storms are usually accompanied by torrential rain.

In the vicinity of the eye a wind direction shift occurs very quickly, while it may even change by 180 degrees abruptly after a short calm in the eye itself. Special procedures are therefore required in order for wind turbines to survive hurricane conditions. Tropical storms are also accompanied by very high turbulence intensity at high wind speeds, an unpredictable situation that is very destructive for wind turbine structures. Therefore, they are usually treated as force majeure in a sales contract of wind turbines.

9.5 Meso-scale Wind Systems

Meso-scale winds encompass atmospheric processes that possess characteristic horizontal dimensions ranging from typically a couple of to a few hundred kilometres and time scales of a few minutes to less than 24 hours. They are the intermediate scales between the synoptic scale and the micro scale. The adaption of large scale winds into tangible local wind processes mainly happens at the meso scale, which makes it important for wind power applications. Physical mechanisms with the strongest impact at the meso scale, which are less important for larger scales, are friction of the air with the surface, local radiative fluxes at the surface, that is, heating and cooling of the ground, turbulent heat and moisture fluxes out of and into the ground, and terrain heterogeneity, like a land–sea temperature contrast and roughness step changes. Mechanical forcings, such as channelling and blocking of flows by complex topographies, though important for meso-scale processes, will be presented in the next chapter, noting that most atmospheric processes in the boundary layer are of meso scale in time and space. Instead, this section will focus on interpreting local thermal forcings on meso-scale winds.

9.5.1 Convection and Thunderstorms

The sun warms the ground, which in turn warms the air directly above it. The warmer air expands, becoming lighter than the surrounding air mass, and starts to rise due to the positive buoyancy force. As the rising air accelerates, it cools due to its expansion at lower high-altitude pressures. Acceleration continues until the air has cooled to the same temperature as the surrounding air. The rising air, if having enough momentum, will continue to rise until negative buoyancy decelerates the parcel to a stop. This is called free convection. Associated with the rising air is a downward flow surrounding the convective column. The downward moving exterior is caused by colder air being displaced at the top of the colume. Below the column, the air mass will start to converge and eventually enter and contribute to the existing convection. Converging air near the surface can cause strong winds in front of the thunderstorm.

For dry air, free convection is only possible in an unstable atmosphere, that is one in which the vertical lapse rate is below the adiabatic one. For moist air, falling temperature due to the rising movement causes some of the water vapour to condense. Condensation releases additional energy (latent heat), which allows the rising air to cool more slowly than if the parcel of air was dry, continuing the ascension. The environment can in this case be statically stable for dry air yet unstable for moist air. With sufficient instability and moisture in the atmosphere, this process will continue long enough for cumulonimbus clouds to form, which support lightning and thunder, namely thunderstorms. Storms typically occur in groups aligned along the front and then move along approximately parallel to the direction of travel of the front. Several such convective storms may combine into a larger system of cells with common dynamics.

All thunderstorms, regardless of type, go through three stages: the developing stage, the mature stage and the dissipation stage. The developing stage is marked by a cumulus cloud being pushed upward by a rising column of air (updraft). The cumulus cloud soon looks like a tower as the updraft continues to develop. There is little to no rain during this stage but occasional lightning. The thunderstorm goes into the mature stage when precipitation begins to fall out of the storm, creating a column of air pushing downward (downdraft). The rain-cooled downdraft spreads out along the ground and forms a gust front, or a line of gusty winds. The mature stage is when hail, heavy rain, frequent lightning, strong winds and tornadoes are most likely to happen. Hailstones usually travel up and down through the convective columns several times, all the time while growing. They only fall out of the cloud towards the ground when their falling speed becomes larger than the vertical speed in the updrafts. In order to grow hailstones of a few centimetres in diameter, the vertical speed in the cloud must reach up to 50 m/s. Eventually, a large amount of precipitation is produced and the updraft is overcome by the downdraft, beginning the dissipating stage. At the ground, the gust front moves out for a long distance from the storm and cuts off the warm moist air that was feeding the thunderstorm. Rainfall decreases in intensity, but lightning remains a danger [8].

Thunderstorms are of scales of several tens of kilometres and therefore are classified as meso-scale convective systems. They produce strong winds that cannot be used for wind energy generation, such as tornados, downburst or thunderstorm gusts. Lighting strikes are also a hazard brought about by thunderstorms.

9.5.1.1 Tornados

Under particularly favourable conditions, the rotation of one or more thunderstorms can lead to tornados. Offshore tornados are called watersprouts. Tornados are a rare occurrence and generally short-lived (typically no more than an hour), but are very destructive on their path, typically a hundred kilometrres wide and up to several tens of kilometres long. The winds felt on the ground are a combination of the translational speed and the rotational speed, together reaching up to 100 m/s. Large tornados may contain subsidiary vortices, which travel around the main vortex, creating a complex pattern of destruction. Being so destructive for wind turbines, the probability of their occurrence and their possible strength should be investigated during wind turbine siting.

Over 1000 tornadoes are recorded in the USA every year, while they are not uncommon in northern Europe, eastern Asia and Australia. It has been estimated that for the German North Sea, an offshore wind park of an area of about 100 km^2 (10 × 10 km) would expect one watersprout within a hundred years [9]. This is already an alarming probability as the sizes of offshore wind parks grow. For example, 3980 km^2 would be required to install 40 GW of offshore wind power in northern Europe, meaning that at least one watersprout would have to be expected every three years.

9.5.1.2 Downburst

A downburst is created by a column of sinking cool air in a thunderstorm that, after hitting ground level, spreads out in all directions. It is capable of producing damaging gusty straight-line winds of over 70 m/s. It starts with calm winds followed by gusty winds with temperature drop. Downbursts blow down trees, flatten crops and wreck buildings. Based on size, a downburst is classified as either a macro-burst, which affects a path longer than 4 km, or a micro-burst, which affects a path of shorter than 4 km. A micro-burst is shorter-lived, but more destructive than a macro-burst.

9.5.1.3 Lightning

Lightning is another threat originating from thunderstorms. It is essentially a giant spark induced when the insulation of air is overcome by the potential difference, charging a massive flow of electrons from the cloud towards the ground. The lightning stroke raises the temperature in narrow channels of air so fast that the air has no time to expand, resulting in shock waves we hear as thunder. Ground to cloud or cloud to ground lightning is actually very erratic in finding contact points. However, due to their exposure, wind turbines and met masts are welcome targets.

9.5.2 Land and Sea Breezes

Another convection-driven weather effect is the land and sea breeze. In fact, meso-scale forcings are most typically induced by pressure gradients set forth by uneven local surface heating, producing local winds that are sometimes very distinct from those on a large scale. These local wind systems are often of such regularity and of sufficient depth and strength that they can be exploited for wind energy generation.

9.5.2.1 Sea Breeze

During the day, the sun heats up both the ocean surface and the land. However, the sea surface heats up much more slowly than the land surface because of the much larger heat capacity of water and because solar radiation penetrates into a thick layer of water, meaning that the mass of water to heat is much larger than that of soil. The air in contact with the land will become warmer compared to the air above the sea surface. This causes the air above the land to rise, creating low pressure at the land surface. Over the water, however, a high surface pressure will form because of the colder air. The wind will blow from the higher pressure over the water to the lower pressure over the land, causing a sea breeze. At a certain height above the surface, usually at around 1–2 km, the rising air above the land blows towards the sea to close the circulation (see Figure 9.10).

A sea breeze front is a weather front created by a sea breeze, also known as a convergence zone. If the air is humid and unstable, the front may create a chain of cumulus clouds and even trigger thunderstorms, propagating up to a hundred kilometres (and even more) inland during the day. The structure of a sea breeze can be quite complicated if the coastline is not straight. The sea breeze strength varies depending on the temperature difference between the land and the ocean. Therefore, it usually appears during the morning and becomes strongest in the late afternoon, fading as the sun sets; it is also typically stronger during the spring and early summer, when the sea is still cold but the sun is strong enough to heat the land surface. Also, sea breezes are weaker in tropical regions due to the milder temperature contrast between the land and the sea.

9.5.2.2 Land Breeze

At night, the roles reverse. The air over the ocean is now warmer than the air over the land. The land loses heat quickly through thermal radiation after the sun goes down and the air above it cools. Because of the much larger heat capacity, however, the ocean is able to hold onto the heat

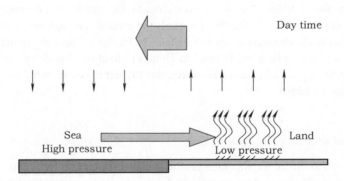

Figure 9.10 An illustration of a fully developed sea breeze circulation

after the sun sets. If the land surface cools below the adjacent sea surface temperature, the air will start to flow offshore, while the returning flow aloft flows in the onshore direction, creating the land breeze. With sufficient moisture and instability, the land breeze can cause rains or even thunderstorms, over the water. But because the marine air is usually stable, such convection is more unlikely. The strength of the land breeze is usually weaker than the sea breeze.

9.5.3   Mountain and Valley Winds

Mountain–valley winds (or mountain–plain winds) are also locally induced, diurnally changing wind systems – like land and sea breezes – that do not emerge from large scale pressure systems but from local temperature differences on the earth's surface (see Figure 9.11).

During the day, the air over the mountain slope heats up more than the air at the foot of the mountain due to greater exposure to solar radiation. This temperature difference creates a relative low pressure at the top of the mountain and high pressure below, forcing a cool breeze to move upwards. To compensate for the upslope flow of air, a breeze on the bottom of the valley (inward valley wind) or on the surface of the surrounding plain (mountain-ward plain wind) is created. It is very common during warmer months when there is a lot of heating from the sun. At night, the air at the upper slope of the mountain cools off much quicker than it does below. The denser air will start to sink and the reverse mechanism applies. At this time, the air is forced to move out of the valley, creating an outward valley wind (also called a mountain wind). Mountain winds are more common in the colder months when there is less warming from the sun. In general, the daytime in-valley winds are stronger and more turbulent than the nightime out-valley winds [4].

Mountain-valley winds are of a spatial scale of a few hundred metres up to a few hundred kilometres, depending on the dimension of the valley. Mountian–plain winds have the largest scale of a few tens of kilometres to over one hundred kilometres [4]. They usually take a few

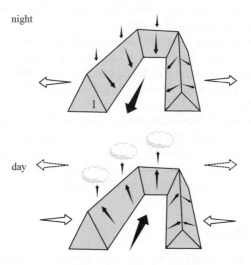

Figure 9.11   An illustration of fully developed mountain–valley and mountain–plain winds at daytime (bottom) and at night-time (top)

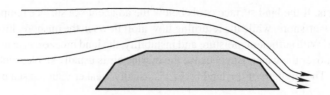

Figure 9.12 Illustration of katabatic winds

hours to develop fully. Their development requires clear skies so that surface heating by the sun and cooling by thermal radiation can occur efficiently. By and large, they are weaker systems of winds than the land and sea breezes. Sometimes, it is even difficult to differentiate them from the synoptic scale motions.

Mountain–valley and mountain–plain winds owe their existence to slope. Slope winds are the winds that blow along the slope (upslope during the day and downslope at night), initiating the circulation of mountain–valley or mountain–plain winds. Due to their shallow depth, however, slope winds are of little interest to wind energy generation. Daily upslope winds tend to blow in a thicker layer because of some convective mixing, while the night-time downslope winds can be really shallow.

9.5.4 Katabatic Winds

A katabatic wind, the technical name for a drainage wind, is a wind that carries high-density air from a higher elevation down a slope under the force of gravity (see Figure 9.12). It is also a purely thermally induced orographic flow, which has similarity with slope winds. Radiational cooling of the air atop a plateau, a mountain, glacier or even a hill forms a temperature inversion in the atmospheric surface layer. The heavier cool air will flow downwards, warming adiabatically as it descends. The temperature of the wind therefore depends on the temperature in the source region and the amount of descent.

Katabatic winds can rush down elevated slopes easily reaching hurricane speeds, though most are not as intense as that. Such intense katabatic winds can also be called downslope windstorms. The accompanying strong turbulence and gustiness adds to their violent character. Sites located downstream of the mountain in the prevailing wind direction are particularly in danger of possible downslope windstorms.

Examples of katabatic winds include the bora (or bura) in the Adriatic, the bohemian wind in the Ore Mountains, the mistral, the Santa Ana in southern California, the tramontane and the oroshi in Japan. The entire near-surface wind field over Antarctica and Greenland is largely determined by the katabatic winds, particularly outside the summer season, owing to doomed topography and radiative cooling of the snow surface [10].

9.6 Micro-scale Winds

A micro-scale wind is short-lived (lasting only a few minutes at a maximum) and of the extent of about 1 km or less. The line between micro-scale and meso-scale winds are not clear-cut and they are sometimes grouped together to study all phenomena smaller than the synoptic scale.

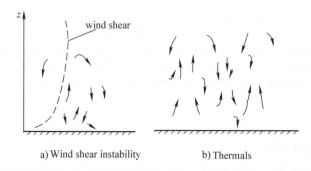

a) Wind shear instability   b) Thermals

Figure 9.13   Main turbulence generation mechanisms in the atmospheric boundary layer

From a wind resource assessment perspective, micro-scale studies show variations of the wind within a wind park area. Micro-scale winds, from the meteorological point of view, occur on the bottom of the atmospheric boundary layer, that is the surface layer, where exchange processes of energy, gases, etc., between the atmosphere and the surface become important. Near-ground turbulent fluxes are thus the main mechanism for micro-scale winds [11]. Apparently the study of micro-scale winds and processes is an essential part of the study of boundary layer winds, which is to be presented in the subsequent chapter. In fact, boundary layer meteorology and micrometerology are often considered to be synonymous [2].

Micro-scale winds generally can no longer be studied as defined systems. They are ususaly described by time-averaged wind speeds and direction (mean wind) and their fluctuations (turbulence).

9.6.1   Turbulence Kinetic Energy

Turbulence kinetic energy (TKE) has been briefly mentioned in Section 3.1.2 to introduce wind flow modelling. Just as kinetic energy is one-half of the product of mass and velocity squared, TKE is based on the squares of the variations in velocities of all three components. The kinetic energy of turbulence per unit of mass can be written as

$$TKE = \frac{1}{2}(u'^2 + v'^2 + w'^2) \tag{9.17}$$

where u', v' and w' denote the turbulence term of three components. Typical values of TKE range from 0.05 m^2/s^2 at night to 4 m^2/s^2 or greater during the day.

Turbulence in the atmospheric boundary layer can be generated by buoyant thermals and by mechanical eddies (see Figure 9.13).[3] Mechanical generation is a consequence of wind shear instability in regions, while buoyancy production is essentially associated with local pockets of unstable air. Turbulence is suppressed by statically stable lapse rates and is dissipated into internal energy (heat) due to the effects of molecular viscosity. Turbulence dissipation is the transport of kinetic energy to smaller and smaller scales, until vanishing into heat.

---

[3] Waves are another turbulence generation mechnisam in the atmosphere. Mountain waves and shearing gravity waves are caused by gravity. Another occurs when two layers of air move with different speeds over each other. In the shear/friction zone a wave will occur and develop into turbulence.

TKE, usually the direct product of wind flow modelling, reflects turbulence in boundary layer flows, but its translation to statistical turbulence intensity as we commonly know it in wind data analysis is not that straightforward. The main challenge to this is that cup anemometers measure a horizontal wind speed and the vertical contribution of turbulence is ignored. The vertical component can be very strong, particularly during the day, and is likely to affect wind turbine productivity.

9.6.2   Turbulent Flux

In computational fluid dynamics (CFD), turbulent flux correponds to the Reynolds stress terms in the RANS equations (see Section 3.1.2). Turbulent fluxes, caused by turbulent mixing, can be visualised as irregular swirls of motion called eddies, which are of many different sizes superimposed on top of each other. The largest eddies in the atmospheric boundary layer have sizes roughly equal to the depth of the boundary layer, that is 100 to 3000 metres in diameter. Smaller eddies feed on the larger ones. The dissipating effects of molecular viscosity becomes stronger for the smallest eddies, which are of the size of a few millimetres. The study of turbulent flux provides quantatitive knowledge of atmospheric processes involving turbulence, which is a prerequisite for the description of boundary layer winds. A mathematical description of turbulence flux is outside the scope of this book, but important physical implications of turbulence flux for understanding micro-scale boundary layer winds are presented here.

9.6.2.1   Taylor's Hypothesis and Advection

In micrometeorology, flux usually refers to the vertical transport of a quantity (e.g. mass, heat, moisture, momentum and pollutant) by eddies. Moving air will cause local changes of the quantity in question by bringing different values from the neighbourhood. This is a processss called advection. When eddies are involved, it is difficult to obtain an instantaneous snapshot of the boundary layer and all eddies formed within it. Taylor's hypothesis assumes that turbulent eddies can be considered to be 'frozen' as it advects, and therefore the observed fluctuations over time at a fixed point are associated with the advection of turbulence transported by the mean wind. It should be noted that Taylor's hypothesis is only valid where the turbulent eddies evolve with a timescale larger than the time it takes the eddy to be advected past the fixed point.

9.6.2.2   Surface Energy Balance

Energy is exchanged between the atmosphere and the earth's surface by turbulent mixing. Solar radiation provides energy to warm the surface. Some of this energy is stored in the ground or oceans. Some of it is returned to the atmosphere by longwave radiation of the surface warming the air and by the latent heat due to evaporation. Inevitably, there is a large diurnal variation in this energy balance over the land surface. As demonstrated in Figure 9.14, after sunrise there is a continuous receipt of downward shortwave radiation from the sun, usually resulting in a positive net heat flux and thus heating of the land surface. Over large bodies of water, the diurnal variation is much less due to the large heat capcity of water and the large depth of medium absorbing the solar radiation.

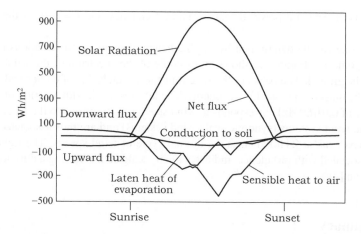

Figure 9.14 An illustration of the surface energy balance over land

9.6.2.3 Turbulent Mixing

This surface energy balance is the driver for forming the structure of the atmospheric boundary layer. Bouyant thermal plumes are created when the surface temperature is higher than that of the air above it, forming a vertically well-mixed layer extending from the surface layer to the top of the boundary layer. This is commonly the case during sunny days over relatively dry land surfaces. Turbulent mixing is reduced as a result of the radiative cooling (sensible heat from the surface to the air) of the surface becoming dominant after sunset. Accordingly, the boundary layer takes a new stable form and tends to shrink in depth. The diurnal variations of the boundary layer in terms of structure and atmospheric stability will be presented in the next chapter.

9.6.3 Turbulence Spectra

When turbulenct flux is visualised as eddies, turbulence spectra (or power spectra) can be defined by the relative strengths of these differentscales of eddies, indicating the contribution of each different size of eddy to the total turbulence kinetic energy. According to Taylor's hypothesis, each frequency is associated with a different size of turbulent eddy. The standard deviation of the wind components are integral values over the entire turbulence spectra. The frequency dependence of the standard deviations needed for wind turbine load calculations can only be obtained from turbulence spectra [4].

For a neutrally stratified atmosphere over a flat and homogeneous terrain, the three-dimensional turbulence spectra fuctions are given by Kaimal *et al.* (1972) [12] as

$$S_u(n) = 105 \frac{z u_*^2}{U(1 + 33f)^{5/3}}$$

$$S_v(n) = 17 \frac{z u_*^2}{U(1 + 33f)^{5/3}} \tag{9.18}$$

$$S_w(n) = 2 \frac{z u_*^2}{1 + 5.3f^{5/3}}$$

where $S(n)$ is the spectral power density, n is the frequency and $f = nz/U$ is the normalised frequency.

The eddy size can be determined by the frequency of wind-speed variations (again using Taylor's hypothesis). Figure 9.2 shows an example of the spectrum of the wind speed measured near the ground. The peaks in the spectrum show which size of eddies and scale winds contribute the most to the total kinetic energy. The leftmost peak with a period of about 100 h (a frequency of 0.01/60 Hz) corresponds to wind speed variations due to the passage of fronts and synoptic wind systems (synoptic scale wind systems) and are thus associated with mean flow. The next peak at 24 h indicates the diurnal cycle of wind speeds. The rightmost peak, which is associated with turbulence, indicates micro-scale eddies having a duration of around 10 s to 10 min [2].

9.7   Summary

This chapter has presented winds of various scales and their corresponding driving forces and mechanisms. The scale of winds is generally classified into three continuous cascades: synoptic scale, meso scale and micro scale. The wind we feel at a certain spot is the result of all scales of wind superimposed on top of each other. Large scales of wind sytems (or eddies) break into smaller and smaller ones, until dissipating into heat.

The forces acting on an air parcel to accelerate consist of a pressure gradient force due to the temperature difference, friction from the surface, Coriolis force because of the earth's rotation, centrifugal force if the path is curved and gravitational force, which is only important in the case of severe vertical motion. The pressure gradient force is the only force in the atmosphere actually initiating motion, while the rest (except the gravitational force) is acting to change the path of the motion. The important geostrophic wind, an exact balance between the Coriolis force and the pressure gradient force on horizontal surfaces, is taken as an example of the work of the balance of the forces.

At a global scale, the three-cell circulation acting to transport energy between the equator and the polar regions is introduced. At synoptic scales (one to two thousand kilometres in diatmeter), cyclones, anticyclones and weather fronts are the general mechanisms responsible for conversion of atmospheric potential energy into kinetic energy in mid-latitudes. In tropical areas, such energy conversion is dominanted by tropical storms.

Typical meso-scale winds are a couple of to a few hundred kilometres in the horizonal dimension. Meso-scale forcings are most typically induced by pressure gradients set forth by uneven local surface heating, producing local winds that are sometimes very distinct from those on a large scale. These meso-scale wind systems include convective storms, land and sea breezes, mountain and valley winds and katabatic winds.

Micro-scale winds are generally interpreted as turbulence due to the high-frequency wind speed variations. Turbulence kinetic energy (from an energy perspective), turbulent flux (from a momentum perspective) and turbulence spectra (from a frequency perspective) are thus presented in this chapter. Knowledge of turbulence is particularly important for understanding boundary layer winds, to be introduced in the subsequent chapter.

References

[1] US Standard Atmosphere (1976) *US Government Printing Office*, Washington, D.C., 1976

[2] Stull, R.B. (1988) *An Introduction of Boundary Layer Meteorology*, Springer.

[3] Orlanski, I. (1975) A Rational Subdivision of Scales for Atmospheric Processes. *Bulletin of the American Meteorological Society*, **56**, 529–530.

[4] Emeis, S. (2013) *Wind Energy Meteorology*, Sprincer Verlag Berlin and Heidelberg.

[5] Subsidence Inversions. http://www.maybeck.com/inversions/ (last visited on 5 June 2014).

[6] Weather Front. http://en.wikipedia.org/ (last visited on 5 June 2014).

[7] Frank, W. M. (1977) The Structure and Energetics of the Tropical Cyclone I. Storm Structure. *Monthly Weather Review*, **105** (9), 1119–1135.

[8] Extreme Weather 101. http://www.nssl.noaa.gov/education/svrwx101/, NOAA National Severe Storms Laboratory (last visited on 8 June 2014).

[9] Dotzek, N., Emeis, S., Lefebvre, C., Gerpott, J. (2010) Watersprouts over the North and Baltic Seas: Observations and Climatology, Prediction and Reporting. *Meteorological Zeitschrift*, **19**, 115–129.

[10] Renfrew, I.A., Anderson, P.S. (2006) Profiles of Katabatic Flow in Summer and Winter over Costs Land, Antarctica. *Quarterly Journal of the Royal Meteorological Society*, **132**, 779–802.

[11] American Meteorological Society. http://glossary.ametsoc.org/wiki/Micrometeorology, (last visited on 7 June 2014).

[12] Kaimal, J.C., Wyngaard, J.C., Izumi, Y., Coté, O.R. (1972) Spectral Characteristics of Surface Layer Turbulence. *Quarterly Journal of the Royal Meteorological Society*, **98**, 563–589.

10

Boundary Layer Winds

The atmospheric boundary layer (ABL), also referred to as the planetary boundary layer (PBL), is the lowest layer of the atmosphere, formed as a consequence of the interaction between the atmosphere and earth's surface. This is the layer where we live and where wind energy is exploited.

This chapter will introduce the basic laws for winds in the boundary layer. These laws are important for understanding wind flow modelling and wind resource assessment, particularly in recent years, because the increasingly larger sizes of wind turbines and wind farms have led to much more complicated interactions between the turbines and the lower atmosphere. As this book concentrates on wind energy generation, aspects such as gravity wave and foehn generation will not be addressed in this chapter; more general descriptions of the ABL can be found in Stull (1988) [1] and many other books.

10.1 Atmospheric Stability

This vertical movement of air, either upward or downward, is generally influenced by the degree of stability or instability of the atmosphere at any particular time (see Figure 10.1). Atmospheric stability is a measure of the atmosphere's tendency to encourage or deter vertical motion.

Physically, whether the air is stable or unstable is determined by the temperature difference between an air parcel and the air surrounding it. The term 'adiabatic process' means warming by compression or cooling by expansion without a transfer of heat or mass into the system. As an air parcel moves up or down within the atmosphere, it is affected by this process. As a result, the temperature of the air parcel should decrease at the adiabatic lapse rate as it moves upwards due to adiabatic expansion. The actual environmental lapse rate, however, usually does not match the adiabatic lapse rate, resulting in a temperature difference between the parcel and the surrounding environment. The adiabatic lapse rate of dry air is about $-10\,\mathrm{K/km}$, that is the temperature of a dry air parcel decreases by about 10 degrees K per kilometre as it moves upwards.

Wind Resource Assessment and Micro-siting, Science and Engineering, First Edition. Matthew Huaiquan Zhang.
© 2015 John Wiley & Sons, Ltd. Published 2015 by John Wiley & Sons, Ltd.

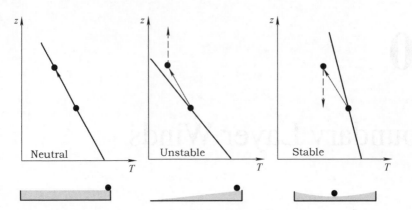

Figure 10.1   Three cases of vertical parcel displacement in neutral (left), unstable (centre) and stable (right) surrounding atmosphere. T denots temperature; z denotes height

10.1.1   Neutral Stratification

Consider a vertical displacement of an air parcel (see Figure 10.1, left). In a neutral stratification, the parcel temperature will decrease by exactly the same amount as the surrounding temperature and because the air pressure adjusts instantaneously, the parcel will also have the same density as the environment. No buoyancy forces will appear to accelerate the parcel and it will remain neutral; that is, a neutrally stratified atmosphere occurs when the actual temperature lapse rate equals the dry adiabatic lapse rate of $-10\,\mathrm{K/km}$. We can imagine a ball on a perfectly flat floor; displacing the ball to any location would not cause further motion. The system is in a state of equilibrium. Because buoyancy is negligible in a neutral stratification, turbulence production is dominated by mechanical eddies.

Although uncommon for the atmosphere the neutral condition is important as the dividing line between stable and unstable conditions. Neutral stability occurs on windy days or when there is cloud cover such that strong heating or cooling of the earth's surface is not occurring. A similar situation also occurs diurnally near sunrise and sunset.

10.1.2   Unstable Stratification

In unstable conditions where the environmental lapse rate is greater than the adiabatic one, a lifted parcel of air will still be warmer than the surrounding air. Because it is warmer and ligher, positive buoyancy forces will appear and cause the parcel to accelerate vertically (also see Figure 10.1, centre), that is the initial vertical moment of the parcel is encouraged. Buoyancy will increase with further vertical displacement.

In the ABL, the vertical movement of the air starts with surface heating. The initial lift is provided by warming of an air parcel over a warmer surface, for example a hot parking lot and an exposed soil surface during a clear day. The parcel, once lifted, will start rising faster and faster and the initial displacement will grow. This is an unstable situation like a ball on a sloping surface; once the rolling starts it will keep going. The degree of instability depends on the degree of difference between the environmental and adiabatic lapse rates, or more straightforwardly for wind resource engineering, on the temperature of the earth's surface, and a higher surface temperature means more instability.

As the air rises, compensating cooler air moves underneath. It, in turn, may be heated by the earth's surface and begin to rise. Under such conditions, vertical motion in both the upward and downward directions is enhanced and considerable vertical mixing occurs.

Unstable conditions usually develop on sunny days when the surface is heated considerably by solar radiation and with low wind speeds so that vertical motion becomes more prominant. Another condition that may lead to instability is the cyclone (low-pressure system), which is characterized by rising air, clouds and precipitation. The earth's atmosphere is generally considered to be unstable overall and as a result the weather is subject to a high degree of variability through distance and time.

10.1.3 Stable Stratification

When the environmental lapse rate is less than the adiabatic lapse rate of -10 K/km, the lifted parcel will have a temperature lower than the surrounding air and negative buoyancy forces will appear to return the parcel to the original height, resisting vertical motion (again see Figure 10.1, right). We can still use the motion of a ball as an analogy, where this time the ball is placed at the centre of a curved trough. The ball will return to the centre and stabilise again after displacement.

Stable stratification is a result of cooling at the surface during night-time. Cooling can be caused by radiative loss of energy after sunset, when outgoing long-wave radiation flux is larger than the incoming solar radiation flux. Wind speed is usually low near the surface. Another instance is when warm air flows over a cold surface; cooling therefore occurs near the surface. This commonly occurs in offshore wind situations when warm air flows over a cold ocean.

Very stable stratification occurs when the temperature increases with an increase in height. This is a 'plus' temperature lapse rate, or an inversion. A layer of inversion acts like a cap, resisting vertical motion of the air below while isolating the air above from surface friction. This can happen when the earth's surface cools rapidly. If the air near the surface cools to a temperature below that of the air above, an inversion called radiation inversion will appear, creating a sharp contrast in wind speeds below and above the inversion layer. Radiation inversions usually occur in the late evening through to the early morning under clear skies with calm surface winds, when the cooling effect is greatest. In colder inland climates, this inversion can be as low as only a few tens of metres, which is within the sweeping area of model large wind turbines. Other causes of inversions occurring at higher altitudes are of little relevance to wind energy generation.

10.1.4 Stability Parameter

Atmospheric stability profoundly affects turbulent mixing in the boundary layer. Unstable conditions will promote turbulent mixing driven by buoyancy. As the surface temperature becomes cooler than the air above it during the night, thermally stable conditions take over and the heat flux reverses sign. Near dawn and dusk, the atmosphere transitions through neutral stability, wherein mechanical shear dominates turbulence production and buoyancy remains negligible. The stabliliy parameter, ζ, is used to characterise the relative influences of buoyancy and mechanical shear. It is given by:

$$\zeta = \frac{z}{L} \tag{10.1}$$

where z denotes the height above the surface and L represents the Monin–Obukhov length [2] and

$$L = \frac{\theta_v}{\kappa g} \frac{u_*^3}{\overline{\theta_v' w'}} \tag{10.2}$$

Here the overline represents a suitable time average, a prime indicates a fluctuating quantity with the 'mean' removed, g denotes the gravitational constant, κ represents the von Kármán constant of magnitude approximately 0.4, w represents the vertical velocity and θ_v is the virtual potential temperature,[1] which includes the modifying influence of the humidity on atmospheric stability. Therefore, $\overline{\theta_v' w'}$ is the nonvanishing virtual potential heat flux at the ground.

The friction velocity, u_*, can be written as

$$u_* = \sqrt{-\overline{u'w'}} = \sqrt{\frac{\tau_w}{\rho}} \tag{10.3}$$

where τ_w denotes surface shear stress or turbulent momentum flux, that is the strength of the momentum sink caused by the presence of the surface, and ρ denotes air density.

Friction velocity, u_*, is always positive in the surface layer. By definition, L is usually negative in the daytime because $\overline{\theta_v' w'}$ is typically positive (upward heat flux) over land, positive at night when $\overline{\theta_v' w'}$ is typically negative (downward heat flux) and becomes infinite at dawn and dusk when $\overline{\theta_v' w'}$ passes through zero. During the day, L is the height at which the buoyant production of TKE is equal to that produced by the mechanical shearing action of the wind [2]. Consequently, the dimensionless stability parameter, ζ, is negative for unstable stratification, positive for stable statificaion and zero (equal to $L \rightarrow \infty$) for neutral stratification.

10.1.5 Modification on a Vertical Wind Profile

In Section 2.1.2 we saw that for a thermally neutral atmosphere and a flat terrain, a vertical wind profile in the atmospheric surface layer is in the logarithmic form of

$$u(z) = \frac{u_*}{\kappa} \ln\left(\frac{z}{z_0}\right) \tag{10.4}$$

This logarithmic equation is deduced on the assumption that turbulence eddies close to the surface in neutral atmospheric conditions have a circular shape, with their size (mixing length) l depending on the distance from the surface z, that is $l = \kappa z$. Also, the turbulent momentum flux is constant in the surface layer, so that the following relation can be formed:

$$\frac{\partial u}{\partial z} = \frac{u_*}{l} = \frac{u_*}{\kappa z} \tag{10.5}$$

In a non-neutral atmosphere, however, the shape of the turbulent eddies is not round and isotropic. They can be visualised as a horizontally elongated, even pancake-shaped feature in

---

[1] Potential temperature, θ, is an artificial temperature that stays constant during vertical displacement of an air parcel without the condensation process.

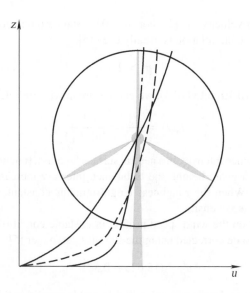

Figure 10.2   The influence of atmospheric stability on a wind profile in the surface layer. Dashed line: neutral; solid line: stable; dot-and-dashed line: unstable

a stable atmosphere and vertically elongated, even in columns, in an unstable atmosphere. A correction to the logarithmic profile has to be made for a satisfactory theoretical description of the wind profile in the surface layer (see Figure 10.2), because correct estimation of shear is critical for vertical extrapolation of wind speeds.

To correct the logarithmic wind profile, a stability function, Ψ_m, is introduced to Equation (10.5):

$$u(z) = \frac{u_*}{\kappa} \ln\left(\frac{z}{z_0} + \Psi_m(\zeta)\right) \tag{10.6}$$

For unstable stratification [3]

$$\Psi_m(\zeta) = 2\ln\left(\frac{1+X}{2}\right) + \ln\left(\frac{1+X^2}{2}\right) - 2\arctan(X) + \frac{\pi}{2} \tag{10.7}$$

where

$$X = (1 - 16\zeta)^{1/4} \tag{10.8}$$

In stable conditions, the correcting stability function for the logarithmic wind profile reads (for $0 < \zeta \leq 0.5$) [4] as

$$\Psi_m(\zeta) = -5\zeta \tag{10.9}$$

10.1.6   Influence on Turbulence

In neutrally stratified atmospheric conditions, the constant turbulent flux assumption for the surface also implies that the standard deviations of the wind speed are independent of height and scale with friction velocity, u_* [1]. For the 10 Hz standard deviations of the streamwise wind velocity, the following relation is usually used [5]:

$$\sigma_u = 2.5u_* \tag{10.10}$$

Inserting Equation (10.10) into (10.4), we get the streamwise turbulence intensity:

$$I_u(z) = \frac{\sigma_u}{u(z)} = \frac{1}{\ln(z/z_0)} \tag{10.11}$$

Therefore, the turbulence intensity in a neutral surface layer is a function of height and roughness length only. For a given height, the turbulence intensity increases with the roughness length of the surface. When the roughness length remains constant, increasing height will result in lower turbulence intensities.

Similar to the effect on the wind speed profile, in unstable conditions in the surface layer, Equation (10.10) has to be corrected using the stability parameter [5]:

$$\sigma_u = (15.625 - 0.5\zeta)^{1/3}u_* \tag{10.12}$$

Due to the sink of turbulent momentum flux in stable conditions, turbulence intensity is usually very low in the surface layer.

10.2   Orographic Effects

Many of the orographic effects on the wind in neutral atmospheric conditions have been discussed analytically in Chapter 2. This chapter explains the interaction between the wind and the terrain in a more theoretical and general manner. This becomes important for a better understanding of wind flow modelling and estimating its uncertainties in various orographic and atmospheric conditions, because the horizontal and vertical adaptions of the wind to orography is so different from one case to another.

This section focuses only on mechanical modifications of the existing larger-scale winds by the underlying orographical features as thermal effects due to uneven local heating of the surface have been introduced in the previous chapter.

10.2.1   Channelling of Wind

Channelling effects of the terrain on the wind flow are generally of meso-scale phenomena. The most prominent result of channelling is wind direction change due to the great constraint from the terrain.

Channelling can be loosely put into two categaries: forced channelling (see Figure 10.3) and pressure-driven channelling (see Figure 10.4). Forced channelling (or deflection) changes the wind direction close to the side walls of the mountain ridges. These side walls cause frictional differences in the normal direction, forcing the wind to align preferentially along the side wall and the valley axis. Pressure-driven channelling occurs when the wind is forced through a narrower path (e.g. a valley), usually resulting in higher wind speeds. The direction and speed of the channelled wind depends on the sign and magnitude of the component of larger-scale winds

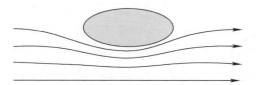

Figure 10.3   Sketch of forced channelling of the wind

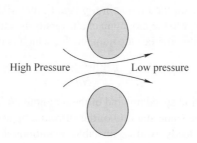

Figure 10.4   Sketch of pressure-driven channelling of the wind

relative to the valley axis and the details of the terrain. Channelling can thus be a favourable condition for wind energy generation, especially when the prevailing wind direction is close to the direction of the valley axis. The abundancy of wind resource in Taiwan Strait, for example, owes much to the channelling effects created by the mainland and Taiwan Island.

Smaller-scale winds may pass through narrow passages in mountain ranges, creating gap winds. Though similar to the channelling effect in terms of large wind speeds, gap winds usually exhibit high turbulence as well due to their small spatial scales. Gap winds can be a frequent phenomenon when the prevailing large-scale wind direction is perpendicular to large mountain ranges. Special attention should be paid to exploiting gap winds during wind turbine micro-siting, and on-site measurement with met masts or ground-based remote sensing are strongly recommended in order to access the specific flow features.

10.2.2   Wind Speed-up and the Froude Number

If channelling occurs when the wind flows around mountains, speed-up effects should reflect modification of the wind flowing over the mountains. The ability to determine whether flow impinging on a mountain will rise over the mountain or will be blocked and deflected around the mountain, or in a combination of both, and what the speed-up (if any) will be is of great practical importance in regions of complex terrain and requires a more detailed analysis. Chapter 2 has presented the speed-up effects of terrain for neutral atmospheric conditions with relatively small height varations, where the speed-up is governed by the horizontal and vertical dimensions of the hills. However, in reality wind speed and atmospheric stability are two other crucial parameters on which the wind regime will depend.

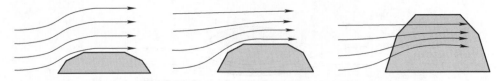

Figure 10.5   Unstable (left) and relatively stable air flow over (centre) and around (right) a mountain of various heights

The parameter of whether the flow can make it over the mountains or not and whether it will accelerate or decelerate on the mountain top (see Figure 10.5) is basically a ratio of the horizontal wind perpendicular to the mountain chain versus the atmospheric stability times the mountain height. This is called the Froude number, F_r, which is of the form:

$$F_r = \frac{U}{Nh} \tag{10.13}$$

where U is the horizontal wind speed upwind of the mountain, h is the mountain height and N is a parameter describing the static stability of the surrounding atmosphere.

A larger Froude number leads to more unstable conditions and $F_r \to \infty$ for neutral and prone unstable conditions. Thus, the Froude number is only suitable for a stable stratified atmosphere. In general, stable stratified atmospheric conditions have the effect of opposing vertical displacement of the air and therefore favour speed-up on the mountain top.

Figure 10.6 illustrates three cases of the wind flowing over a mountain in stable atmospheric conditions. In very stable conditions with the presence of temperature inversion (not shown in Figure 10.6) and at low wind speeds, the impinging wind is more likely to flow around the mountain and may even be completely blocked by the mountain if the inversion layer is lower than the height of the mountain. A unique cloud pattern (called a 'von Kármán vortex street') formed by complete blocking of the flow in very stable conditions is shown in Figure 10.7.

At slightly stronger wind speeds and in less stable conditions (see Figure 10.6, left) at least part of the wind will flow over the mountain. The air, once being pushed over the mountain by the wind, tends to get restored into its original height by gravity. The result is that these wave-like oscillations move away from their origin. The properties of these waves (length, amplitude, speed and direction) are set by the environment and the way they are created. With enough moisture you may even see the interesting cloud patten shown in Figure 10.8, when the air condenses due to the rising motion [6].

When $F_r \approx 1$, the atmosphere will become more unstable with higher wind speeds (see Figure 10.6, centre). The air can easily climb over, creating gravity waves large enough to be comparable to the mountain in dimension. Vortexes and downslope windstorms can occur on the leeside.

Figure 10.6   Winds flow over a mountain in three different prone-stable conditions. PE represents potential energy and KE represents kinetic energy

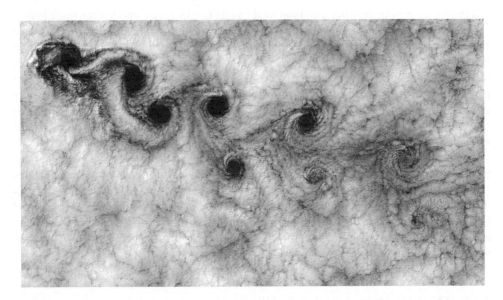

Figure 10.7   NASA satellite image (Landsat 7) of clouds off the Chilean coast near the Juan Fernandez Islands on 15 September 1999

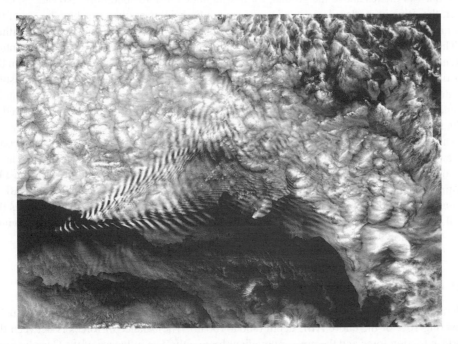

Figure 10.8   NASA satellite image (MODIS imager on board the Terra satellite) of a wave cloud forming off Amsterdam Island in the far southern Indian Ocean. Image taken on 19 December 2005

A further increase in the wind speed and weakening of atmospheric stability can result in flow separations on the leeside and reversed wind direction, as shown in Figure 10.6 (right). When $F_r \to \infty$ or $F_r \gg 1$, in the case of very strong wind speeds and neutral condions, the wind will be disturbed long before it reaches the mountain. Within the influenced zone, which is about three times the dimensions of the mountain, the wind will flow through a virtually narrower path usually capped by an inversion layer on top of the boundary layer. This inversion basically creates a possibility for the Venturi effect by not allowing the whole boundary layer to be lifted and thus slowed down, consequently causing some speed-up. Otherwise, neutral or unstable air should be more likely to cause wind to decelerate while traversing a mountain.

In summary, wind speed-up is stronger in more stable conditions, except for very stable stratification, which may completely block the flow. The dimension and orientation of the mountains also determine the air–terrain interactions, and so does the wind speed. These complex interations become more important when it comes to meso-scale flow modelling. Micro-scale wind flow modelling tools that are used in wind resource assessment resolve the flow based on the assumption of flow similarity in a relatively small area, thus extrapolating the winds in the atmospheric surface layer.

10.3   Onshore Boundary Layer Winds

So far, we have been mostly concentrating on the winds in the atmospheric surface layer, which is about 100 metres above the ground, for mid-latitude onshore conditions. Nowadays, however, the siting of modern large wind turbines with tip heights easily exceeding the top of the surface layer inevitably requires more detailed understanding of the ABL and the winds within.

The vertical structure and extent of the ABL depends largely on the energy and momentum exchange between the surface and the atmosphere above it. Its complexity originates from the spatial and temporal variations of the surface properties, including shape, coverage, moisture content, heat capacity and radiation fluxes, etc. For simplicity, assuming a flat, horizontally homogeneous terrain in an equilibrium state, the vertical layering of the ABL is as depicted in Figure 10.9.

The viscous roughness sublayer directly above the surface is generally ignored, as it is only a few millimetres deep and has no relevance to wind energy generation. The surface layer (also called the Prandtl layer or constant-flux layer) and the Ekman layer of the ABL must be distinguished due to the difference in the balance of forces within them. Together, the ABL is about 300 to 3000 m in depth, capped by a layer of temperature inversion that separates the ABL from the free troposphere above. The free troposphere is where the geostrophic winds (see Subsection 9.3.1) prevail.

10.3.1   Surface Layer

The atmospheric surface layer is the layer where air is in direct contact with the surface, usually covering 10% of the entire ABL depth (about 100 m). The balance of forces is between the surface friction force and the pressure gradient force only. The influence of the Coriolis force is negligible in this layer and therefore the wind direction does not change with height. Vertical turbulent fluxes of momentum, heat and moisture are considered to be constant in the surface layer, so that the theoretical logarithmic wind profile can stand as valid. This is why the surface

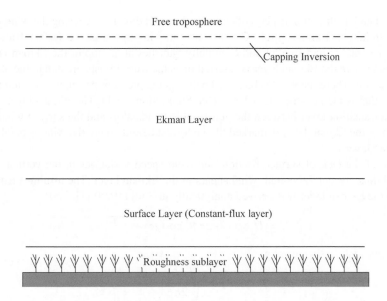

Free troposphere

Capping Inversion

Ekman Layer

Surface Layer (Constant-flux layer)

Roughness sublayer

Figure 10.9   Vertical structure of the atmospheric boundary layer over flat and homogeneous terrain

layer is also referred to as the constant-flux layer. Turbulence intensity at a certain height within the surface layer is only dependent on surface roughness. Theories and engineering applications of wind resource assessment introduced in previous sections and chapters are by and large based on the physics of the surface layer.

For terrain with gentle height variations, the depth of the surface layer can somewhat follow the terrain surface. Over steeper terrain, such as hills and mountains, the summits can easily exceed the surface layer depth and be exposed to different wind regimes.

Forest-covered surfaces are special in that the depth of the roughness sublayer becomes substantial. In fact, only the crowns of dense forests form a rough surface in common with grass lands from which the constant-flux layer and logarithmic wind profile starts. Below the crowns is the canopy sublayer (about two-thirds of the canopy height), which is virtually a substitute for the real earth's surface, presenting a displacement height (see Subsection 2.1.5) to vertical profiles of the wind. In the case of relatively sparse forests, the wind can easily penetrate the canopy sublayer, feeding the surface layer directly above the canopies with an addtional high momentum turbulent flux. This layer of high turbulence intensity may extend to approximately three to five tree heights. Therefore, wind turbines at forest sites should have hub heights at least three times the tree height in order to avoid this anomalous high turbulence [5]. Urban surfaces are similar to sparsely vegetated forests with respect to the canopy sublayer and the anomalous high turubence above and within.

10.3.2   Ekman Layer

Moving up from the surface layer into the Ekman layer, which covers about 90% of the whole ABL depth, the Coriolis force (see Section 9.2) due to the earth's rotation becomes important in the force equilibrium as well. The magnitude of the Coriolis force is proportional to the wind speed and its direction is perpendicular to the wind direction, causing the wind direction to turn with height (Ekman spiral). Continuously moving upwards, the surface friction force will become smaller and smaller and eventually vanishes at the top of the Ekman layer. The troposphere above the Ekman layer is referred to as the free atmosphere due to the absence of surface friction. There the wind is balanced between the pressure gradient force and the Coriolis force, that is the geostrophic balance (see Subsection 9.3.1). The Ekman layer therefore acts like a transition layer between the geostrophic winds aloft and the surface winds below. The depth of the Ekman layer is marked by the lowest height where the wind is parallel to the geostrophic layer.

Because of the lack of surface friction, the wind speed variations in the vertical are very small and thus present very small wind shears in the Ekman layer. The turning vertical wind profile in the Ekman layer was derived analytically in Stull (1988) [1]:

$$U = U_g\sqrt{(1 - 2e^{-\gamma z}\cos(\gamma z) + e^{-2\gamma z})} \tag{10.14}$$

The length scale, γ, is given by

$$\gamma = \sqrt{\frac{f}{2K_m}} \tag{10.15}$$

where U_g denotes the geographic wind vector and f denotes the Coriolis parameter. K_m, the turbulent-transfer coefficient, is asummed to be constant in the Ekman layer of neutral or stable boundary layer.

The Ekman layer depth, h_E, is defined as

$$h_E = \frac{\pi}{\gamma} \tag{10.16}$$

Equation (10.14) should not be extrapolated down into the surface layer where K_m cannot be assumed constant. Works have been done to unify the vertical wind profile for the boundary layer. Readers can refer to Emeis (2013) [5] for a detailed description. Generally speaking, the unified wind profiles depend on surface roughness, geostrophic wind speed, surface layer height, friction velocity and the angle between the surface wind and the geostrophic wind.

10.3.3   Diurnal Variations

Diurnal changes in incoming radiation and thus surface temperature govern the structural changes in the ABL. The depth of ABL is highly variable due to entrainment at the top and rising thermal mixing from the surface, as shown in Figure 10.10.

In clear sky daytime conditions (statistically unstable), the surface is heated by the sun, causing convective plumes to rise. These plumes expand adiabatically as they rise until a thermodynamic equilibrium is reached at the top of the atmospheric boundary layer. The overshooting (updrafts) of these plumes and sinking (downdrafts) of drier, wamer air from the free troposphere at the top of the layer is called entrainment. Due to intense turbulent mixing, heat, moisture and momentum are nearly constant in the vertical, although strong wind shear can occur in the surface layer due to friction. This well-mixed boundary layer is called the convective boundary layer (CBL). Near sunset, radiative cooling occurs on the surface, allowing turbulence to decay slowly in the formerly well-mixed CBL. The resulting neutrally

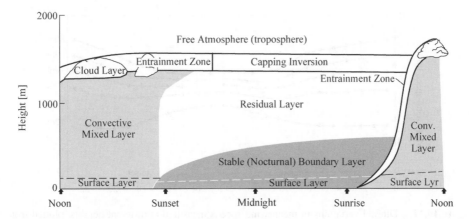

Figure 10.10   Dirunal cycle of boundary layer structure in clear sky conditions over land [1] *Source*: Springer Science+Business Media (adapted)

stratified layer of air is called the residual layer (RL). The RL cools slowly and more-or-less uniformly during the night. As the night progresses, the bottom portion of the residual layer is transformed into a stable boundary layer (SBL) due to direct contact with the rapidly cooling ground, forming an inversion layer that separates the RL from the surface. The depth of the SBL grows during the night, until approximately sunrise, when the surface is heated and convective mixing begins again [1].

These nonstationary features in the diurnal development of the ABL structure have a profound impact on the winds. We should be aware that the wind profiles introduced previously are only for stationary conditions or mean conditions over a long period of time. The SBL decouples the winds above and below the SBL height. The winds above therefore accelerate due to the missing frictional force from below, while the winds below slow down because they no longer feel the driving winds from higher layers. This development process of the winds at different heights is illustrated in Figure 10.11. The plots of mean daytime and night-time vertical profiles cross each other at a height known as the crossover height, where a minimum diurnal variation of wind speed can be found [5]. The crossover height is closely related to the depth of the nocturnal SBL, which varies from tens to hundreds of metres mainly depending on the cooling efficiency of the surface.

10.3.4   Low-Level Jets

The development of nocturnal SBL leads to a sudden disappearance of the dragging frictional force in the force equilibrium on the winds. Wind speed thus shoots to much higher values and a new wind maximum in the profile is born, as shown in Figure 10.12. This phenomenon is called a low-level jet. Similar to the force balance in the Ekman layer, the Coriolis force provokes a turning of the wind vector as well.

The peak velocity in low-level jets is typically about a hundred metres or so above the surface. High wind speeds very close to the surface can also appear when the ground surface is extremely cold and the atmosphere near the surface is extremely stable with very strong

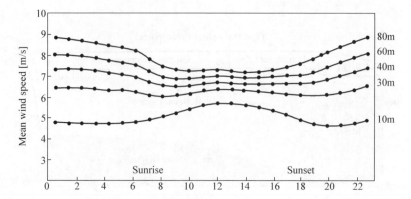

Figure 10.11   Diurnal variations in mean wind speeds measured at different heights, plotted from a year of wind data from a met mast located in Northest China

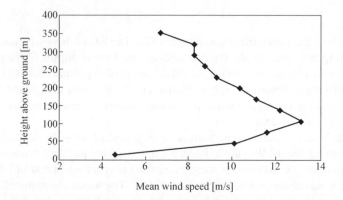

Figure 10.12   An example of an observed wind profile in the presence of a nocturnal low-level jet

temperature inversion. In northern China, Russia and Canada, for example, low-level jets may appear at heights well below 100 m, which is lower than the tip height of modern wind turbines.

Over land, the development of low-level jets usually requires clear skies, so that thermal stratification near the surface can change rapidly and a sudden transition from an unstable daytime CBL to a nocturnal SBL can occur. In this case, the remaining strength of the horizontal synoptic pressure gradient determines the wind speeds in low-level jets. Low-level jets can also occur when warm air suddenly crosses over to a cold, smooth surface, such as from warm land to a colder ocean surface or from bare land to snow or ice-coverd surfaces.

Low-level jets, if they happen below the wind turbine tip height, can increase energy production of the wind turbine at night due to increased wind speeds within its rotor area. The setbacks are the possible intense turbulence generated as a result of inertial oscillation of the horizontal wind speed and strong mechanical shear. Ground-based remote sensing tools might be necessary to assess their impact on wind turbine production and loads if wind profile patterns resembling Figure 10.11 or Figure 10.12 are identified or where low-level jets are suspected to happen below the wind turbine tip height.

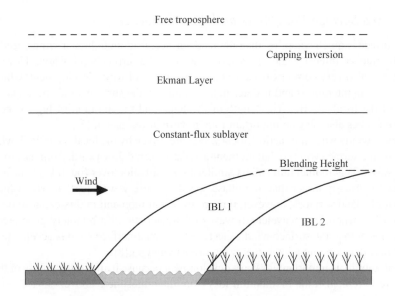

Figure 10.13   Sketch of developing IBL over a step-change in surface property

10.3.5   Internal Boundary Layer

Horizontal advection of air across a discontinuity in roughness or thermal of the surface leads to the formation of an internal boundary layer (IBL) [7]. It grows in height with the distance from the line of step change until a new equilibrium is reached and the IBL is blended with the rest of the boundary layer (see Figure 10.13). The IBL and its engineering implications in wind resource assessment have been introduced in Section 2.1.3. Here, the IBL will be briefly presented as modifications to the boundary layer structure.

As the height of the internal boundary layer grows, a new wind profile corresponding to the new surface property is developed downstream. In general, the wind measured at a mast at a given height is representative of the surface properties in an upstream distance of about 100 times the measurement height [5].

10.4   Offshore Boundary Layer Winds

The first offshore wind power project was installed off the coast of Denmark in 1991. Since then, commercial-scale offshore wind power facilities have been operating in shallow waters of countries such as the United Kingdom, Germany, Norway, the Netherlands, Japan, China, South Korea, Belgium, Sweden, Italy, Portugal and others. The offshore wind resource is very abundant. The wind is strong and consistent in comparison with onshore conditions. We can thus expect that an increasingly larger portion of wind energy will be generated at offshore wind farms in the future. This section will present the characteristics of offshore winds in the marine atmospheric boundary layer (MABL) and applications of offshore wind resource assessment and wind turbine layout design.

10.4.1   Sea Surface Roughness and Wave Influence

The sea surface is much smoother than the land surface, leading to higher wind speeds, much smaller turbulence intensity and less wind shear over the turbine's rotor area. However, the sea surface is also very complex because of the dynamics of wave development, which is the result of air–sea interations and is determined mainly by the surface wind speed and thermal stability of the local MABL. The length of the fetch and the duration of high wind speeds over these fetches also play an important role in wave development [5].

In terms of wind–wave interaction, local waves are driven by the local wind field, which acts as a drag on the surface wind so that momentum is transferred downward, from the atmosphere into the waves. This momentum sink is similar to the situation over rough land surfaces. The wind speed is usually larger than the phase speed of these waves, particularly at high wind speeds. Unlike land surfaces, however, upward turbulent momentum flux can also occur over the sea surface when long wavelength waves (characteristic of a remotely generated swell) propagate faster than the surface wind. As a result, the near-surface wind is accelerated in this case and may develop into a low-level wave-driven wind jet [8].

Thermal stability is important for wave development because it affects the ability of the lower atmosphere to replenish the momentum loss from the atmosphere above, or vice versa. This being true, the waves are expected to be higher for more unstable conditions, as also is the sea surface roughess length, whereas the onshore surface roughness is assumed to be a function of surface characteristics only and is independent of atmospheric conditons such as wind speed and thermal stability.

The sea surface roughness is on the order of a tenth of a millimetre to a millimetre for moderate wind speeds relevant to the operating wind speeds of wind turbines (see Section 2.1.1). The sea surface roughness length can become several decades higher in the case of high and extreme wind speed events, because of the induced much larger wave heights. Disregarding viscosity and surface tension of the water surface, Charnock (1955) [9] proposes an empirical function that describes the dependence of water surface roughness on the shear production of turbulence:

$$z_0 = \alpha \frac{u_*^2}{g} \tag{10.17}$$

where g is the gravitational acceleration, u_* the friction velocity and constant α denotes the Charnock parameter. The value of the constant α is usually quoted as 0.01–0.04, where the lowest value is for the open sea and the highest value is for shallow and near-coastal conditions [10]. The IEC standard 61400-3 (2006) [11] assumes $\alpha = 0.011$ for offshore conditions.

Taking the logarithmic wind profile of Equation (10.6) into (10.17), we get

$$z_0 = \frac{\alpha}{g} \left(\frac{\kappa u(z)}{\ln\left(\frac{z}{z_0} - \Psi(\zeta)\right)} \right)^2 \tag{10.18}$$

Equation (10.17) is the Charnock relation corrected by surface-layer stability.

The discussion on the roughness length of the sea surface should distinguish smooth flow and rough flow. For low wind speeds on the open ocean, the thickness of laminar flow can cover the roughness elements (ripples) and therefore the flow is considered smooth and the roughness length increases with the wind speed. As the wind speed increases further, the roughness length of the surface of large water bodies, such as the ocean, depends on the fluctuation of waves and the flow is thereby called rough flow, which Equation (10.17) is designed for. It is also assumed that ocean waves belong to the family of gravitational waves and that wave heights depend on the intensity of wind shear in the atmospheric surface layer. This means that the roughness length of an open water surface increases rapidly with the wind speed.

10.4.2   Marine Atmospheric Stability

The sea surface is an infinite source of moisture. Turbulent heat fluxes in the marine surface layer depend on the temperature difference between the sea surface and the air immediately above. They are directed upwards when the sea is warmer than the air and downwards for the opposite situation. Because the air above is nearly always drier than the air directly over the sea surface, turbulent moisture fluxes nearly always point upwards. Since air density slightly decreases with increasing humidity, the air directly over the sea surface will tend to rise, leading to a slightly unstable marine surface layer. For example, it was observed that within a long marine fetch, the conditions in the Dutch North Sea and Danish North Sea are dominated by unstable and near-neutral conditions [12]. Neglecting the humidity influence therefore may lead to a biased wind profile and thus incorrect vertical extrapolation of the mean wind speed [13].

Thermal stability is determined by the air–surface temperature difference. Due to the very large heat capacity of the ocean, the sea surface temperature does not present diurnal variations. Instead, it stays almost constant for 24 hours. Therefore, the air–sea temperature difference in the short term is created not because of the change in solar radiation but as a result of the temperature contrast between the sea surface and the advected air. When cold air is advected over a warm sea surface, the thermal condition is unstable; it is stable in the case of warm air advections over a cold sea surface. In mid-latitude, the offshore air temperature change in the course of a few days is often a result of moving atmospheric pressure systems and the development of weather fronts. The cold and warm air advections are coupled to different wind directions because of the prevailing westerly winds. Therefore, we can expect that in the temperate lattitudes of the northern hemisphere, the north-westerly and northerly winds are likely to be unstably stratified (cold air advection) and south-westerly winds are likely to be stable (warm air advection). In contrast, thermal stability has no correlation with the wind direction over land, where it is overwhelmed by diurnal variations of land surface temperature induced by radiative fluxes.

10.4.3   Annual and Diurnal Variations

As introduced previously, thermal properties of the surface are decisive for the development of a boundary layer structure. Instead of diurnal variations seen on land, the open sea surface temperature presents a dominant annual variation. A maximum temperature appears in later summer and a minimum in late winter. Therefore, an unstable marine boundary layer prevails in autumn and early winter because the atmospheric temperature cools more rapidly than the sea and cooler air masses over the warmer sea water are found. Similarly, stable offshore conditions prevail in spring and early summer as the atmosphere rapidly warms. The thermal inertia of the sea water with respect to the atmosphere is responsible for the seasonal pattern of MABL development. This annual variation is slightly modified by cold and warm air advections induced by moving atmospheric pressure systems on a temporal scale of a few days [5].

Near the coast, the situation is more complex due to the land and sea breezes. A diurnal variation pattern can be found. The wind that blows from the land at night-time has a lower temperature than the sea surface, leading to cold air advections and thus stable stratification. When the sea surface is considerably colder than the adjacent land surface and the wind is directed from the land to the sea by a larger scale pressure system, low-level jets can occur as well in the marine boundary layer without the diurnal variation in thermal stability of the surface layer required over land [14]. Near-coast low-level jets are also the result of decoupling between the warm air advection and the surface friction.

10.4.4 Offshore Turbulence Intensity

According to Equation (10.11), offshore turbulence intensity in the atmospheric surface layer depends on surface roughness and is thus a function of wind speed as well. Over the sea surface, mechanical shear turbulence production becomes stronger and stronger with increasing surface wind speed due to higher waves (higher roughness) and becomes dominant when the wind speed reaches a certain high value. Consequently, for high wind speeds of above 11–13 m/s, offshore turbulence intensity increases near linearly with increasing wind speed, unlike the usual, monotonically decreasing turbulence intensity curve (plotted against wind speed) for onshore situations [5]. Minimum offshore turbulence intensity appears at wind speeds of around 10–12 m/s, which are roughly the starting point of the rated power of wind turbines. The combination of increasing wind speeds and increasing turbulence intensities is a characteristic load case for offshore wind turbines and should be carefully investigated during the load calculation.

Besides wind speed, studies (e.g. [5] and [15]) based on the data from the two research platforms in the German Bight FINO1 and FINO3 have shown that offshore turbulence intensity is also dependent on direction, despite the homogeneous sea surface. This is fundamentally due to the direction-dependent characteristics of marine atmospheric stability. We have already discussed previously that thermal stability has a direct influence on turbulence intensity as it is the indicator for buoyant turbulence generation. Offshore, this effect is amplified by the development of waves, which are also strongly affected by thermal stability of the atmosphere.

10.4.5 Offshore Vertical Wind Profile

Because of the smaller turbulence flux, the offshore surface layer or constant-flux layer is often much shallower than that onshore. The depth can be in the order of merely 10 m for stable conditions and low to moderate wind speeds [5]. As a result, the hub height of an offshore wind turbine and much of the rotor area are more likely to be merged in the Ekman layer, where, as mentioned before in Section 10.4.1, the wind speed increases only slightly with height. Therefore, increasing the hub height of offshore wind turbines usually does not make much economical sense, as doing so only ends up with a very small increase in energy production. In the Ekman layer, the wind direction also turns with height. However, this turning of the wind vector is usually of little consequence within the height of the turbine rotor.

The smooth surface makes the power law wind profile more suitable for offshore winds than it is for onshore winds [5] (see Subsection 2.1.6). The wind shear (power law exponent) values for offshore winds are much lower than those over land. The standard IEC 61400-3 (2006) [11] for offshore wind turbines assumes in the normal wind profile (NWP) model that the wind shear is below 0.14 for neutral offshore conditions. This wind shear value can be easily exceeded for stable atmospheric conditions. The stability-dependent property of the offshore wind shear component is expected to be in the same manner as that for onshore. What makes the offshore wind shear unique is the fact that it also increases considerably with growing wind speed due to the resulting rougher sea surface. Offshore wind shear also depends on wind direction, again due to the direction-dependent characteristics of marine atmospheric stability.

10.4.6 Offshore Turbine Layout Optimisation

In contrast to onshore sites, the offshore 'terrain' is flat and uniform without spatial changes in roughness, so that identical free wind conditions over the size of common offshore wind farms can be a good assumption. This being true, offshore wind turbine micro-siting should be an easier task than onshore, but then it is not without its unique engineering challenges. Foundation cost (function of water depth and geological environment), power cable cost (function of the distance between turbines and between the farm and the main grid connection point) and logistics for construction and maintenance are some of the major challenges facing the engineers and offshore wind power project developers. These factors should be considered while making decisions on the placement of offshore wind turbines. However, for the scope of this book, we will focus on wind resource assessment only.

The free wind conditions for offshore wind farms are relatively simple and typically uniform. The actual difficulties thus arise not from the site but from the wind farm itself, that is turbine wakes. In fact, reducing wake effects is the main task for offshore wind turbine layout optimisation. Offshore turbulent momentum mixing is considerably lower than that onshore due to the low roughness of the sea surface. As a result, turbine wakes are able to propagate much longer downstream before merging into the free stream (see Section 4.3). Secondly, typical offshore wind farms are large in size and equipped with huge turbines (rotor diameter of 120 m and more). For these reasons, the spacings between offshore wind turbines should be substantially larger than onshore. Another contributor to turbulent mixing is surface layer stability, which for offshore is correlated with wind direction. The wind direction with prevailing warm air advections (stable) should have even larger distances between wind turbines. Therefore, the relation between average atmospheric stability and wind direction should be analysed during wind resource assessment for offshore wind farms. Two identical temperature sensors should be mounted at two different heights of the met mast so that such an analysis can be perfomed. In general, automated layout optimisation tools are more capable for offshore wind parks than they are for onshore sites given the simpler free wind conditions and homogeneous 'terrain' condtions.

When it comes to production uncertainty analysis (see Chapter 7) for offshore wind parks, one should bear in mind that most offshore turbine power curves are still measured onshore with different atmospheric conditions in terms of turbulence characteristics and vertical wind profiles, etc. These differences are particularly important because of the large rotor sizes of offshore turbines. Therefore, additional uncertainties should be introduced if this is the case.

10.5   Summary

Atmospheric thermal stability is critical for the development of ABL structure and the interaction between the winds and the terrain. The atmospheric surface layer is thermally neutral when the actual lapse rate is equal to the dry adiabatic lapse rate, unstable when the actual lapse rate is larger and stable when it is less. Thermal stability over land depends mainly on the net radiative flux on the ground and presents a clear diurnal cycle, whereas the sea surface temperature does not change during the day due to the much larger heat capacity. The marine atmospheric stability is formed by advections of the air with different temperatures from the sea surface. Cold air advections lead to unstable offshore conditions and warm air advections lead to stable conditions.

Whether the flow impinging on a mountain will rise over the mountain or will be blocked and deflected around the mountain, or in a combination of both, and whether there is wind speed-up depends on the atmospheric stability, the wind speed and the horizontal and vertical dimensions of the mountain. Wind speed-up on a mountain top is generally stronger in more stable conditions, except for very stable stratifications, which may completely block the flow.

The atmospheric boundary layer is classified into three layers according to different balances of forces. The roughness sublayer directly above the surface is of no relevance to wind energy applications. The surface layer, also called the constant-flux layer, is where the logarithmic wind profile is considered valid. Turbulence intensity in the surface layer is only a function of surface roughness in neutrally stratified conditions. The Ekman layer above occupies about 90% of the atmospheric depth. The Coriolis force is no longer negligible in the Ekman layer, leading to a turning of the wind vector with height. The wind vector becomes parallel to the geostrophic layer on top of the Ekman layer. The marine surface layer is very shallow due to the low shear turbulence generation and thus offshore wind turbines generally operate in the Ekman layer. Low-level jets are formed due to the sudden loss of frictional force in the force equilibrium. They can appear both over land and over the sea (near the coast).

The development of waves on the sea surface is a function of wind speed. This means that the roughness of the sea surface is also dependent on wind speed. Consequently, offshore turbulence intensity increases with growing wind speed at high wind speeds of over $11-13$ m/s. The distinctive features of offshore winds are the directional dependency of thermal stability, the turbulence intensity and the vertical wind profile. Offshore wind turbine layouts should adapt to this direction-dependent atmospheric stability because it has a strong impact on turbine wakes.

References

[1] Stull, R.B. (1988) *An Introduction of Boundary Layer Meteorology*, Springer.
[2] Obukhov, A.M. (1971) Turbulence in an Atmosphere with a Non-uniform Temperature (English Translation). *Boundary-Layer Meteorology*, **2**, 7–29.
[3] Paulson, C.A. (1970) The Mathematical Representation of Wind Speed and Temperature Profiles in the Unstable Atmospheric Surface Layer. *Journal of Applied Meteorology*, **9**, 857–861.
[4] Dyer, A.J. (1974) A Review of Flux-Profile Relations. *Boundary Layer Meteorology*, **1**, 363–372.
[5] Emeis, S. (2013) *Wind Energy Meteorology*, Springer Verlag, Berlin and Heidelberg.

[6] Image of Wave Clouds. http://earthobservatory.nasa.gov/IOTD/view.php?id=6151 (last visited on 8 July 2014).

[7] Garratt, J.R. (1990) The Internal Boundary Layer – A Review. *Boundary-Layer Meteorology*, **50**, 171–203.

[8] Hanley, K.E. Belcher, S.E. (2010) A Global Climatology of Wind–Wave Interation. American Meteorology Society, USA.

[9] Charnock, H. (1955) Wind Stress on a Water Surface. *Quarterly Journal of the Royal Meteorological Society*, **81**, 639–640.

[10] Petersen, E.L., Mortensen, N.G., *et al.* (1997) Wind Power Meteorology. Risø National Laboratory, Denmark.

[11] IEC 61400-3 (2006) Wind Turbines – Part 3: Design Requirements for Offshore Wind Turbines.

[12] Sathe, A., Gryning, S.-E., Peña, A. (2011) Comparison of the Atmospheric Stability and Wind Profiles at Two Wind Farm Sites over a Long Marine Fetch in the North Sea. *Wind Energy*, **14**, 767–780. doi: 10.1002/we.456.

[13] Barthelmie, R.J., Sempreviva, A.M., Pryor, S.C. (2010) The Influence of Humidity Fluxes on Offshore Wind Speed Profiles. *Annals of Geophysics*, **28**, 1043–1052.

[14] Brooks, M., Rogers, D. (2000) Aircraft Observations of the Mean and Turbulent Structure of a Shallow Boundary Layer over the Persian Gulf. *Boundary-Layer Meteorology*, **95**, 44–50.

[15] Westerhellweg, A., Canadillas, B., Neumann, T. (2010) Direction Dependency of Offshore Turbulence Intensity in the German Bight. *10th Wind Energy Conference DEWEK*, Germany.

11

Environmental Impact Assessment

Harnessing power from the wind is one of the cleanest and most sustainable ways to generate electricity. It is commonly accepted that the development of renewable energy sources and especially wind energy is one of the most viable solutions to the problem of climate change, one of the greatest threats to the future of mankind. For these reasons, we have seen a spectacular increase in installed wind power capacity in recent years. Despite the vast potential, the very rapid development of wind power facilities was accompanied, inevitably, by concern in local communities over their possible impacts on the environment. In some cases, the objections to wind power installations are valid to some extent and deserve further investigation, so that the potential environmental impacts can be recognised and mitigated. However, in many other cases, the worries conveyed seem somewhat exaggerated and sometimes unrealistic.

The environmental concerns of wind power plants have been raised over the noise produced by the rotor blades, visual impacts and deaths of birds and bats that fly into the rotors. Wind resource engineers are often asked to quantify some of these impacts so that the local communities can be objectively informed and cases argued on a factual basis. Sometimes, compromises have to be made on the positioning of wind turbines in order to mitigate the impacts, which may lead to a reduction in power production or an increase in installation costs. This is why these and other environmental concerns associated with wind energy development are discussed in this chapter.

11.1 Biological Impacts

The impacts of wind turbines on wildlife, most notably on birds and bats, have been widely documented and studied. Appropriate wind farm siting and best management practice are recommended by many governments and organisations in order to mitigate these impacts and deliver more biological-friendly wind power projects. In the majority of cases, these impacts could be minimised, to the level where they are of no significant concern, through careful siting.

The Environmental Impact Assessment (EIA) Directive 85/337/EEC on assessment of certain public and private projects on the environment (as amended by Directive 97/11/EC and 03/35/EC) and the Strategic Environmental Assessment (SEA) Directive 2001/42/EC on the assessment of the effects of certain plans and programmes on the environment are the two key

Wind Resource Assessment and Micro-siting, Science and Engineering, First Edition. Matthew Huaiquan Zhang.
© 2015 John Wiley & Sons, Ltd. Published 2015 by John Wiley & Sons, Ltd.

pieces of EU environmental legislation that are directly relevant to wind farm developments. Any wind farm development that is likely to affect one or more Natura 2000 sites has to undergo a step-by-step appropriate assessment procedure and, where necessary, apply the relevant safeguards for the species and habitat types of community interest, as explained in the guidance document published by the European Commission in 2010 [1]. In the United States, the Fish and Wildlife Services [2] has played a leadership role in making comprehensive recommendations on appropriate wind farm siting and best management practices.

11.1.1   Birds and Bats

Wind farms may affect birds and bats through direct mortality or lethal injury from collision, displacement and habitat loss due to disturbance and barrier effects [1, 3].

Collisions tend to be related mostly with rotors due, at least in part, to high tip-speeds or with other associated infrastructures such as overhead cables, towers, nacelles, power lines and meteorological masts. A majority of the studies have quoted low collision mortality rates per turbine, arguing that flying birds sometimes run into buildings and other permanent structures as well, but this does not necessarily mean that collision mortality caused by wind farms is insignificant, especially in large wind farms consisting of perhaps several hundreds or thousands of turbines. For large, long-lived species of birds (e.g. raptors and golden eagles) with generally low annual productivity and slow maturity, even relatively small increases in mortality rates may be significant for their populations [4]. Careful site selection is therefore needed to minimise fatalities and it is always recommended to try and site the turbine accordingly to limit bat or bird surveys. Guidance from Natural England suggests that wind turbines should be 55 m from woodland and hedge rows when bats or protected birds are present.

The collision risk depends on a range of factors related to bird species, number and behaviour, flight height and type, the stage of the bird's annual cycle, wind and weather conditions, time of the day, topography and the nature of the wind farm itself, including turbine spacing, hub height, rotor diameter and the use of lighting. Susceptible locations for high collision risks are primarily related to topographical bottlenecks [1], for example mountain passes or land-bridges between water-bodies, on migratory flyways or local flight paths and flight corridors between feeding areas and roosting sites. Other susceptible locations are near feeding or resting sites, for example near wetlands and sandbanks, where large numbers of birds can be found, and slopes where soaring birds are attracted to gain lift. Large birds are generally more vulnerable to collisons because of their poor manoeuvrability. The use of aviation and shipping warning lights on turbines may increase the risk of collision by attracting and disorientating birds, particularly in conditions of poor visibility. It is therefore advised to use the minimum number of intermittent flashing white lights of the lowest effective intensity to lower the risk [3]. Drewitt and Langston (2006) [3] give a comprehensive review on the possible causes of high collision risks. Offshore wind turbines can have similar impacts on marine birds, but as with onshore wind turbines, the bird deaths associated with offshore wind turbines are minimal [5].

It has been reported that more than 90% of bats killed within range of wind turbines actually die because of a sudden change in air pressure before or after each blade, which bursts their blood vessels. This is probably because wind turbines attract vast numbers of insects, both at onshore and offshore locations, through the heat radiation from the wind turbine, which in turn bring larger predators within range of the spinning blades. A recent study has shown that the colour of the turbines themselves also has a significant impact on the number of insects that are attracted to it, both during the day and at night [6]. It is found, according to this study, that turbines painted pure white and light grey drew more insects than any other colour apart from yellow and that wind turbines painted purple would dramatically reduce the number of insects, and therefore bats and insectivorous birds, that were brought within range of the spinning blades [6]. Research into wildlife behaviour and advances in wind turbine technology have helped to reduce bird and bat deaths. For example, wildlife biologists have found that bats are most active when wind speeds are low. Using this information, the Bats and Wind Energy Cooperative [7] concluded that keeping wind turbines motionless during times of low wind speeds could reduce bat deaths by more than half without significantly affecting power production.

Disturbance, leading to displacement and then habit loss, can be caused by the presence of wind turbulance through visual, noise and vibration impacts or as a result of activities related to site construction and maintenance. Though few studies are conclusive, direct loss of or damage to habitat resulting from wind farm infrastructure (turbine bases, substations and access roads, etc.) is not generally perceived to be a major concern for birds, except for sensitive habitats or sites of national and international importance for biodiversity [4]. The barrier effect, also a form of disturbance, may occur when wind farms are located along migration routes or flyways, or, at a more local level, along regular linkages between feeding areas and resting or breeding sites [3].

11.1.2   Terrestrial Animals

The footprint of a wind farm involves not only wind turbines and subsations but also a number of other factors such as access roads, cleared areas and power lines, human activity in the area as part of maintenance work, disturbance during construction and dismantling, and improved accessibility to the area, for example outdoor recreation, hunting and leisure traffic [8]. Disturbance, habitat loss and fragmentaion are the main areas of concern on the impacts on terrestrial animals, particularly large mammals, associated with wind farms [9]. For the larger species, the impacts of wind power depend mainly on the network of access roads to the wind turbines [8].

There is still large knowledge gaps in the field, indicating the need for research as well as for efficient environmental monitoring. Studies are needed, for example, on the effects of noise and visual impacts from the turbines and the localisation of new wind power facilities in relation to areas of particular value for ungulates and large predators [8]. Although the knowledge is generally sparse, most studies have shown that direct habitat loss appears to be small for wind power projects. The animals usually reclaim the site over a relatively short period of time and become accustomed to the disturbances from wind power. By contrast, the presence of wind farms could benefit many wildlife species as open land and edges create new browsing areas; roads can facilitate animal movement in the landscape or help animals escaping parasitic insects [8].

11.1.3   Marine Animals

Wind farms located offshore can have positive impacts on fish and other marine wildlife. Some studies suggest that turbines may actually increase fish populations by acting as artificial reefs

[10]. These artificial reefs provide good breeding conditions with a wide selection of food and shelter from currents, encouraging the creation of new habitats.

The most significant negative impact of offshore wind farms is likely to be the noise produced throughout the life of the development, especially during construction. Pile-driving, for example, is a particulary intense noise source and may disrupt the behaviour of dophins and other marine mammals at distances of many kilometres.The nosie can be harmful for dophins because they use echo-location to navigate, find food and avoid predators. This should be considered in site planning and environmental impact assessments. The industry is also developing engineering solutions to reduce the noise levels produced during construction or using alternative construction techniques that generate less noise.

11.1.4   Vegetation

Onshore infrastructure, including access roads, substations and turbine bases, often require removal of the vegetation. The significance of the vegetation loss usually depends on the size of the area disturbed and whether rare or sensitive native plants or plant communities are affected. The loss may be intensified as a result of the growth of invasive, weedy plant species in disturbed areas [11]. Rare plant field surveys may be conducted by a qualified botanist prior to construction, so that mitigation measures such as segregation and storage of topsoil, soil decompaction and topsoil replacement can be applied to avoid or at least minimise the impacts.

11.2   Visual Impacts

Visual impacts are among the most commonly expressed concerns about the development of wind energy projects. Because they must generally be sited in exposed places, wind turbines are often highly visible. Visual impacts in the vicinity of a wind turbine due to shadow flickering of blades can be assessed and thus avoided through careful micro-siting. Scenic and aesthetic issues are by their nature highly subjective. Though proper siting decisions can help to avoid or minimise any aesthetic impacts to the landscape, or blend them into the landscape, informing the concerned communities of the benefits derived from their use, so that they can be more acceptive, can also be an important measure to take.

11.2.1   Shadow Flicker

Shadow flicker occurs when the turbine blades pass in front of the sun and creates flickering shadows on a building or across the landscape (see Figure 11.1). Wind turbine shadow flicker may induce adverse human health effects include annoyance and stress [12]. The flicker effect is a particular concern for people who suffer from photosensitive epilepsy, which affects about one in 4000 people. Shadow flicker can also be a safety concern. For example, it can cause vehicle driver distraction.

Medical research has shown that a flicker rate of three flashes per second or less has a very low risk of inducing a seizure in sensitive individuals. Therefore, planning should ensure the flash frequency does not exceed three per second and the shadows cast by one turbine on

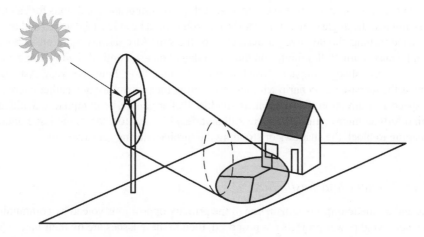

Figure 11.1 Sketch of blade shadow flicker on a residential building

another should not have a cumulative flash rate exceeding three per second [13, 14]. On a typical three-blade wind turbine, this frequency would correspond to a rotation speed of one complete rotation per second (or 60 rpm). However, typical three-blade turbines rotate at much slower rates of 10 to 20 rpm. Therefore, the risk of seizures induced by flickering shadow from unity-scale wind turbines can be considered negligible. The risk is even lower with large modern models due to the relatively slower rotation speeds.

Though severe health impacts due to blade flickers seem unlikely, visual issues raised by close neighbours of wind farm projects are still justified. Fortunately, it is possible to accurately calculate the flicker impacts in terms of frequency, timing and duration at a specific receptor location. Shadow flicker varies by time of the day and season and usually only falls on a single building for a few minutes of a day, depending on the angle of sunlight. This analysis is usually performed through computer-based mapping and modelling modules incorporated in wind resource assessment toolkits such as WindPRO and WindFarmer, using data relating to sun exposure, locations and elevations of both the turbines and the receptors, and turbine parameters such as the hub height, rotor diameter and blade width. More detailed on-site investigation of factors such as tree screening and window orientation, size and tilt can usually determine whether potential impacts indicated in the modelling would in fact occur, and to what extent [11]. For example, in the United Kingdom, only properties within approximately 130 degrees either side of north, relative to the wind turbine, can be affected, as wind turbines do not cast shadows on their southern side. The time of the day and the type of building may also be considered in the evaluation. As an example, an office building would not be affected if all shadow impact occurred after business hours.

Shadow flicker has been a concern especially in Northern Europe where the effect is exacerbated by the high latitude and low sun angle. Many regulatory guidelines have been published regarding this issue. For example, according to the Scottish Government Online Renewables Planning Advice Documents, 'Onshore Wind Turbines', a distance of ten rotor diameters (e.g. 1000 m distance for a wind turbine of 100 m rotor diameter) can be used as the maximum distance that the effect of shadow flicker could extend. The default maximum distance in WindPRO is 2000 m [15]. The shadows can hardly reach the limit due to the optic conditions of the atmosphere. As stated by the German guidelines on limits and conditions for calculating shadow impact, the angle of the sun over the horizon should be at least 3 degrees and the blade of the wind turbine should cover at least 20% of the sun. Also, a maximum of 30 hours per year and a maximum of 30 minutes on the worst day of an astonomical shadow impact can be allowed. If a real shadow impact is used, it must be limited to 8 hours per year. Astonomical shadow is the worst-case scenario without considering the probability of weather overcast and wind speed and direction distribution. If shadow flicker cannot be completely avoided during wind turbine micro-siting, there are still a variety of ways to address the issue, including landscaping to block the shadows or stopping the turbines during sensitive times.

11.2.2   Scenery and Aesthetics

Scenic and aesthetic impacts remain one of the primary oppositions voiced by communities in areas where wind power projects are proposed, though these issues are by their nature highly subjective. To some people, they are graceful sculptures; to others, they are eyesores that compromise the natural landscape. Business owners, and even some environmental groups, are split on how to balance developing renewable power with scenic interests. The size, number, scale, motion and visual prominence of wind turbines makes visual mitigation difficult. One strategy being used to partially offset visual impacts is to site fewer turbines in any one location by using today's larger and more efficient models of wind turbines, but, at the same time, larger wind turbines are generally more visible.

Proper micro-siting decisions can help to some extent to avoid aesthetic impacts to the landscape, but it must also be balanced with a myriad other site constraints, such as wind resource, turbine wake, wildlife impacts, constructability, microwave beam path, property setbacks, landowner preferences and proximity to residences and public roads. Visual issues are by necessity the last criteria by which turbines can be sited [11]. The chapter 'Design as if People Matter: Aesthetic Guidelines for a Wind Power Furture' in Paul Gipe's 2002 book *Wind Power in View* [16] provides a comprehensive discussion on design measures that can be helpful for delivering a more aesthetic wind power project. Some options for key aesthetic design, construction and operation measures are listed in the AWEA *Wind Energy Siting Handbook* [11], including employing uniform turbine units with a balanced shape, colour and size, not using commercial markings or messages on the turbines, limiting the use of lighting and installing underground power collection cables.

Besides aesthetic measures, there are methods to evaluate the scenic conditions and potential impacts of projects, and sophisticated, proven technologies exist for 'viewshed' analysis and visualisations of what projects would look like before they are built. Notwithstanding, there are no general acceptance criteria regarding the maximum allowable visibility of wind turbines. This process is often referred to as the visual impact assessment. For example, the ZVI (zones of visual influence) module in WindPRO calculates and documents the theoretical visibility of wind turbines on the landscape, that is identifying the spots from where one will be able to see one or more turbines [15]. The ZVI method is based on the landscape established from digital height coutours, local obstacles and surfaces (e.g. forest areas). The existing landscape conditions can be characterised through representative sampling of 'before' photographs. The 'after' views can then be superimposed into the 'before' photographs, to the extent that project elements will be visible after construction. This kind of visual simulation is able to provide a basis for characterising the degree of visual contrast that would be created by the project.

11.3 Noise Impacts

Like all mechanical systems, wind turbines produce some noise when they operate. Noise can be defined as any unwanted sounds. Apart from the tonal sounds caused by mechanical components such as the gearbox and generator at the top of the tower, most sounds originate from the flow of air around the blades, mainly during the downwards stroke in front of the tower during a rotation. The so-called aerodynamic noise generally increases with rotor speed and is affected by factors such as the shape of the blade tip, the pitch angle and smoothness of the blade surface. Most of the turbine noise is masked by the sound of the wind itself, and the turbines run only when the wind blows. In recent years, engineers have made design changes to reduce the noise from wind turbines, as noise also reflects lost energy and output. However, wind turbine noise is often deemed to be more annoying than transportation noise because of its repetitive, pulsing nature and high variability in both level and quality – from 'swoosh' to 'thump' to silence, all modulated by the wind speed and direction. In addition, unlike vehicle traffic, which tends to get quieter after dark, turbines can sound louder at night. Also, wind turbines generate lower frequencies of sound, which tend to be judged as more annoying than higher frequencies and are more likely to travel through walls and windows [17].

Because wind power projects are typically located in rural areas where pre-existing masking background sound levels are low, an assessment of potential impacts to neighbours and other sensitive receptors is often required. If necessary, noise levels at nearby residences can be managed through careful siting of turbines and other operational measures.

11.3.1 Wind Turbine Noise Curve

Sounds are characterised by their magnitude (loudness) and frequency (pitch). A sound's loudness is usually measured by the sound pressure level in decibels (dB). Sound pressure or acoustic pressure is the local pressure deviation from the ambient atmospheric pressure caused by a sound wave. The sound pressure level (SPL) is a logarithmic measure of the effective sound pressure of a sound relative to a reference value. It is important to distinguish between the sound power level and the sound pressure level, both of which are the measures of loudness. The sound power level is the property of the sound source and gives the total acoustic power emitted by the source. The sound pressure level is a property of the sound at a given observer location.

Loudness is a subjective measure. Therefore, sound meters are normally fitted with filters (such as A, B or C) adapting the measured sound response to the human sense of sound. The decibel A filter, so-called 'A-weighted', is the most widely used variant. The measurements made using this filter are expressed in units of dB(A). It underweighs low and high pitch sounds and therefore more closely matches how the human ear perceives sound.

Most turbine manufacturers measure and provide turbine sounds following standards set by the IEC (standard 61400-11) [18], which must include sound power levels and dominant frequencies (pitches) at five different wind speeds (6, 7, 8, 9 and 10 m/s) for a specific turbine model. The AWEA (1989) standards [19] and the IEA 1994 recommendations [20] are also widely used to ensure consistent and comparable measurements of wind turbine sound power levels. According to the IEC standard 61400-11, the sound power levels of a wind turbine model should be measured at 10 m height above ground using 8 m/s as the reference wind speed. This does not mean that a measurement has to be made with 8 m/s winds at a height of

10 m. The measurement is typically made using a microphone at ground level at a given distance from the turbine base and then the results are standardised for the reference wind speed and height using a series of mathematical calculations defined by the IEC standards [11]. Specific measurements for infrasounds (lower than 20 Hz), low-frequency noise (20 to 100 Hz), directivity and impulsivity are optional within the current IEC standards. As an example, the noise curves of Vestas V90-2MW wind turbine (active pitch regulated) are given in Table 11.1 and Figure 11.2.

As shown in Figure 11.2, noise from pitch-regulated variable speed wind turbines increases steeply with wind speed and so the rotational speed of blades up to the point at which the tubine is generating its rated power. Above the rated power, there is no increase in noise because the rotational speed is regulated by pitching the blades to maintain a constant power output. For stall-regulated machines, noise can increase considerably above the rated power. It is possible

Table 11.1   V90-2MW noise curve mode 0 (dB) at 1.225 kg/m³

Hub height Wind speed	80 m	95 m	105 m	125 m
4 m/s	94.4	95.0	95.5	96.1
5 m/s	99.4	100.0	100.3	100.8
6 m/s	102.5	102.8	103.0	103.3
7 m/s	103.6	103.7	103.8	103.9
8 m/s	104.0	104.0	104.0	104.0
9 m/s	104.0	104.0	104.0	104.0
10 m/s	104.0	104.0	104.0	104.0
11 m/s	104.0	104.0	104.0	104.0
12 m/s	104.0	104.0	104.0	104.0

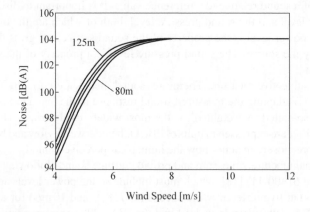

Figure 11.2   V90-2MW noise curves, corresponding to Table 11.1

to reduce turbine noise by derating the rotational speed, but doing so would result in energy production loss.

The ambient background noise tends to increase faster with increasing wind speed than wind turbine noise. As a result, turbine noise issues are more commonly a concern at lower wind speeds, because at wind speeds above 8 m/s it is generally masked by the background wind-generated noise [21].

11.3.2   Sound Propagation

The sound's strength decays in a predictable way as noise travels aways from the wind turbine. Assuming hemispherical nosie propagation over a reflective surface, this can be modelled using the simple propagation equation:

$$L_p = L_w - 10\log_{10}(2\pi R^2) - \alpha R \tag{11.1}$$

where L_p is the sound pressure level (dB) at the receptor, L_w is the sound power level from a noise source (dB), R is the distance from the source (m) and α is the frequency-dependent sound absorption coefficient; $\alpha = 0.005$ dB(A)/m for broadband sounds in the atmosphere.

The simplified Equation (11.1) does not take into account terrain effects (such as blocking and funnelling), ground absorption, wind direction or other atmospheric effects (such as wind speed, temperature gradients and humidity). The discussion of complex propagation models that include all these factors is beyond the scope of this book. These factors are usually modelled by a sound evaluation module offered in commercial wind resource assessment software, following the international rule ISO 9613-2:1996 [22], or country-specific regulations. The ISO 9613-2:1996 describes a method for calculating the attenuation of sound during propagation outdoors in order to predict the levels of environmental noise at a distance from a variety of sources under meteorological conditions [22]. As an example, the noise propagation map of a wind farm consisting of three wind turbines is illustrated in Figure 11.3.

11.3.3   Combining Sound Levels

A wind farm is usually composed of more than one wind turbine, that is multiple sources of noise. Therefore, we need to determine the combined noise levels from the entire wind farm at a certain receptor location. Sound pressure levels cannot be added in a liner fashion, but sound intensity, which is also a measure of sound strength, can. Sound intensity or acoustic intensity is defined as the sound power per unit area in W/m^2. The approximate relation between sound intensity, P, and decibels can be given by

$$P = 10^{\left(\frac{L_p - 90}{10}\right)} \tag{11.2}$$

Thus

$$L_p = 10\log_{10}(P) + 90 \tag{11.3}$$

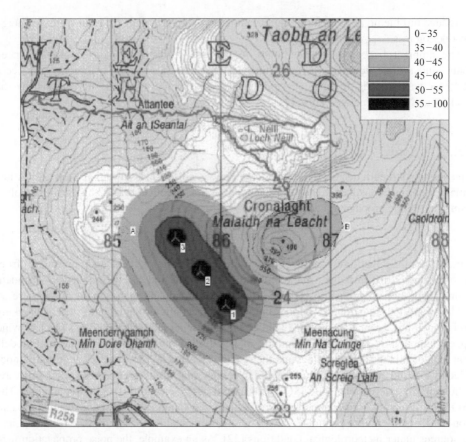

Figure 11.3 Example of a turbine noise propagation map, created with WindPRO [23] *Source*: EMD International A/S

Example 11.1

Two wind turbines with the same sound power level of 103 dB(A) are placed 300 m and 250 m away from the receptor respectively. What is the combined sound pressure level at the receptor's location?

Solution

According to Equation (11.1), the sound pressure level from each individual turbine at the receptor's location can be calculated as

$$L_{p, A} = [103 - 10 \ \log(2\pi \times 300^2) - 0.005 \times 300] = 45.5 \ \text{dB(A)}$$

$$L_{p, B} = [103 - 10 \ \log(2\pi \times 250^2) - 0.005 \times 250] = 49.0 \ \text{dB(A)}$$

Then, following Equation (11.2), the combined sound intensity is

$$P = 10^{\left(\frac{45.5-90}{10}\right)} + 10^{\left(\frac{49.0-90}{10}\right)} = 0.00011 \ \text{W/m}^2$$

Taking the result above to Equation (11.3), we get

$$L_p = [10\log(0.00011) + 90] = 50.6 \text{ dB(A)}$$

The resulting conbined sound pressure level of the two wind turbines is only 1.6 dB(A) higher than the louder noise source.

The addition of two sound levels is illustrated in Figure 11.4. For two values of the same sound level, the total sound level is only 3 dB(A) higher than the individual one. The total sound level becomes closer to the higher one as the difference between the two increases, that is the lower sound level is 'swallowed' or 'masked' by the higher one. Therefore, the arrangement of turbines will affect overall noise levels, but the combined effect may be only slightly louder than a single turbine.

For the same reason, it is often necessary to measure the background baseline noise at a particular location prior to installation of the wind turbines. The relative increase in noise level from the wind farm can be as important as the absolute noise levels of the wind farm itself. The background baseline noise levels will be related to such things as local traffic, industrial and farming machinery, barking dogs, lawnmowers, children playing and, more importantly, the interaction of the wind with gound cover, trees, buildings, power lines, etc. It will vary with the time of day, wind speed and direction and the level of human acivity. In general, a change in the sound level of 1 dB cannot be perceived but a change of 5 dB will typically result in a noticeable community response.

11.3.4   Evaluating Noise Levels

Figure 11.5 is created following Equation (11.1) for a sound power level of 104 dB(A), which is typical for modern MW wind turbines. We know that the noise emitted from a wind turbine already weakens to 40 dB(A) about 400 m away from the turbine base and 40 dB(A) is the limit of night-time noise level recommended by the World Health Organisation to avoid potential health impacts from noise.

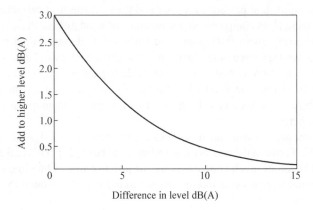

Figure 11.4   Addition of two sound levels

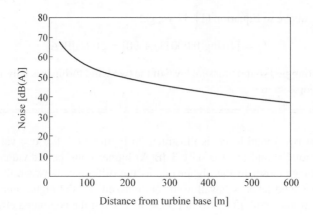

Figure 11.5 Estimated noise levels at different distances from a typical 2 MW wind turbine. The sound power level at the turbine is 104.0 dB(A)

Human perception of sound levels is considerably subjective, as each individual may have a different sensitivity to various types of noise. In some cases, those who are opposed to the wind farm project may be more annoyed by the same sound than those who are in favour of the project [11]. Nevertheless, it is still helpful to compare various sound levels with everyday sounds, as shown in Table 11.2.

Noise regulations depend on the country and are often different for daytime and night-time. Typically, the limits for the wind farm noise level should be 35 to 40 dB(A) for zones for rural living and 40 to 45 dB(A) for other zones. It may be further required not to exceed the background noise by more than 5 dB(A), especially under already nonquiet background conditions.

11.4 Weather and Climate Change

It is thought that the development of wind energy will cut global warming emissions dramatically, therefore saving our planet from the catastrophe of severe climate change. To put this in context, it is estimated that natural gas generated electricity will emit between 0.6 and 2 pounds of carbon dioxide equivalent per kilowatt-hour and the estimate for coal-generated electricity is 1.4 to 3.6 pounds [24]. In comparasion, the estimate of wind turbine life-cycle global warming emissions is merely about 0.03 pounds of carbon dioxide equivalent per kilowatt-hour depending on the wind resource potential of the wind farm. However, large-scale wind power extraction from the atmosphere may have a noticeable impact on regional weather and climate as well, because by converting wind's kinetic energy into electricity, wind turbines modify surface–atmosphere exchanges and the transfer of energy, momentum, mass and moisture within the atmosphere.

A study on a large area of wind farms in Texas, USA, has found a significant local warming trend of up to 0.72 °C over a decade as more turbines are built [25]. This makes sense, since at night the ground becomes much cooler than the air just a couple of hundred meters above the surface and the wind turbines generate turbulent mixing near the ground that causes warmer

Table 11.2 Sound levels associated with everyday sounds and wind turbines in dB(A)

Common sounds	Sound pressure level (dB(A))	Threshold
Jet engine at 30 m	150	
Rifle being fired at 1 metre	140	
Jet taking off 100 m away, pain threshold	130	
Amplified music at 2 m, loud car stereo	120	Hearing damage due to short-term exposure
Roak band indoor	110	
Chain saw, inside subway train	100	
Power tools, lawnmower	90	
	85	Hearing damage due to long-term exposure
Busy traffic at 10 metres, larm clock	80	
Car at 10 metres, vacuum cleaner, commercial area	70	
TV sound at home, dishwasher	60	
Normal conversation	55	(Unitary wind turbine 50 metres away)
Moderate rainfall, quiet urban daytime outdoors,	50	
Quiet urban night-time out-doors	40	WHO recommendated night-time limit (Unitary wind turbine 400 metres away)
Quiet suburban night-time outdoors	35	
A quiet room,whisper, countryside at night	30	
Leaves rustling, calm breathing	10	

air to be transported downwards. As a result, the ground does not get quite as cool as it would otherwise. This same strategy is commonly used by fruit growers who fly helicopters over the orchards to combat early morning frosts.

This warming effect found in the study is generally small and remains a local phenomenon. More challenging is the investigation of global effects resulting from large-scale wind power generation on the order of 10 TW. Wang and Prinn (2010, 2011) [26, 27] studied two such hypothetical scenarios of large-scale wind energy exploitation using atmospheric modelling. They found a surface warming effect of over 1 °C within the wind parks and possible alterations of global distributions of rainfall and clouds, where onshore wind turbines produced 10% of the global demand in 2010 (4.5 TW). In contrast to onshore wind farms, large offshore wind farms were found to cause surface cooling over the installed area, due to the enhanced latent heat flux from the sea surface to the lower atmosphere [28], but overall the disturbance of large-scale deployment of offshore wind turbines to the global climate is relatively small. For this reason, Emeis (2013) concludes that wind energy alone will not be sufficient to meet all of the future energy demands of mankind and other forms of renewable energy need to be considered as well [28].

11.5 Public Health and Safety

Wind turbines do not use combustion of fossil fuels to generate electricity, and hence do not produce air pollution. The only potentially toxic or hazardous materials are relatively small amounts of lubricating oils and hydraulic and insulating fluids. Therefore, contamination of surface or ground water or soils is highly unlikely. General safety issues such as fire and injuries during construction are very similar to any other type of construction work with the presence of heavy equioment.

The primary health and safety considerations for the public are related to blade movement. While cases of blade drop/throw have occurred, these incidents are rare and have generally been linked to improper assembly or extreme windstorms with wind speeds exceeding the design limits. In case of ice build-up, it is possible for the ice on the rotor to shed and, if the rotor is moving, to be cast some distance away. Where icing is expected, it is wise to design the layout with appropriate setbacks from sensitive receptors and educate the operational staff about the conditions likely to cause ice accretion on turbines and the risk of ice falling [11].

An additional concern associated with wind turbines is potential interference with radar and and telecommunication facilities. Like all electrical generating facilities, wind generators produce electric and magnetic fields. There is little evidence that electric fields cause long-term health effects.

11.6 Summary

The astonishing development of wind energy in recent years has showcased its potential as an important energy source but at the same time raised somewhat heated community concerns on its potential environmental impacts. Of those concerns, the most important ones are collisions and disturbance for bird species, shadow flickers of blades and the noise emitted by wind turbines. Through careful siting, most environmental impacts associated with wind enery development can be mitigated.

Bird collision risk depends on a range of factors related to bird species, number and behaviour, flight height and type, the stage of the bird's annual cycle, wind and weather conditions, time of the day, topography and the nature of the wind farm itself, including turbine spacing, hub height, rotor diameter and the use of lighting. Susceptible locations for high collision risks are primarily related to topographical bottlenecks, feeding or resting areas and slopes where the soaring birds are attracted to gain lift. Large birds are generally more vulnerable to collisons because of their poor manoeuvrability. The fatalities of bats and insectivorous birds can be reduced by changing the colour of wind turbines so that fewer insects are attracted to them.

Blade flicker impacts varies by the time of the day and season and usually only falls on a single building for a few minutes of the day, depending on the angle of sunlight. The time of day and the type of building may also be considered in the evaluation. Shadow flicker has been a concern especially in Northern Europe where the effect is exacerbated by the high latitude and low sun angle. A distance of ten rotor diameters is usually used as the maximum distance that the effect of shadow flicker could extend. According to the German guidelines on limits and conditions for calculating the shadow impact, the angle of the sun over the horizon should be at least 3 degrees and the blade of the wind turbine should cover at least 20% of the sun.

Besides careful turbine micro-siting, landscaping and stopping the turbines during the sensitive times can also be used to reduce flicker impacts.

Wind turbine sounds mostly originate from the aerodynamic noise during the downward stroke in front of the tower during a rotation. Most of the turbine noise is masked by the sound of the wind itself, but it is often deemed more annoying than transportation noise because of its repetitive and pulsing nature, as well as the lower sound frequencies. In addition, turbines can sound louder at night due to the lower masking background noise. The propagation and superpositioning of sounds can be calculated by computer codes applying the ISO standards. The arrangement of turbines will affect overall noise levels, but the combined effect may be only slightly louder than a single turbine. In general, a change in sound level of 1 dB cannot be perceived but a change of 5 dB will typically result in a noticeable community response. Human perception of sound levels is considerably subjective, as each individual may have a different sensitivity to various types of noise. The noise of a moden wind turbine can reduce to about 40 dB(A) at a 400 metre distance; 40 dB(A) is the limit of night-time noise level recommended by the World Health Organisation to avoid potential health impacts from noise.

References

[1] EU Guidance on Wind Energy Development in Accordance with the EU Nature Legislation. Available at: http://ec.europa.eu/environment/nature/natura2000/management/docs/Wind_farms.pdf (last accessed 17 July 2014).

[2] Fish and Wildlife Service (FSW) (2010) Recommendations of the Wind Turbine Guidelines Advisory Committee. Available at: http://www.fws.gov/habitatconservation/windpower/Wind_Turbine_Guidelines_Advisory_Committee_Recommendations_Secretary.pdf (last accessed 17 July 2014).

[3] Drewitt, A.L., Langston, R.H.W. (2006) Assessing the Impacts of Wind Farms on Birds. *Ibis*, **148**, 29–42.

[4] Langston, R.H.W., Pullan, J.D. (2003) Windfarms and Birds: An Analysis of the Effects of Wind Farms on Birds, and Guidance on Environmental Assessment Criteria and Site Selection Issues. Report T-PVS/Inf (2003) 12, by BirdLife International to the Council of Europe, Bern Convention on the Conservation of European Wildlife and Natural Habitats and RSPB/BirdLife in the UK.

[5] Hüppop, O., Dierschke, J., Exo, K.-M., Fredrich, E., Hill, R. (2006), Bird Migration Studies and Potential Collision Risk with Offshore Wind Turbines. In Wind, Fire and Water: Renewable Energy and Birds. *Ibis*, **148** (Suppl. 1), 90–109.

[6] Long, C.V., Flint, J.A., Lepper, P.A. (2011) Insect Attraction to Wind Turbines: Does Colour Play a Role. *European Journal of Windlife Research*, **57**, 323–331.

[7] Arnett, E.B., Huso, M.M.P, Hayes, J.P., Schirmacher, M. (2010) Effectiveness of Changing Wind Turbine Cut-in Speed to Reduce Bat Fatalities at Wind Facilities. A Final Report Submitted to the Bats and Wind Energy Cooperative, Austin, TX, USA, Bat Conservation International.

[8] Helldin, J.O., Jung, J., Neumann, W., Olsson, M., Skarin., A., Widemo, F. (2012) The impacts of wind power on terrestrial mammals (English translation report). Naturvårdsverket Report 6499, The Swedish Environmental Protection Agency, Sweden.

[9] Arnett, E.B., Inkley, D. B., Johnson, D.H., Larkin, R.P., Manes, S., Manville, A.M., Mason, J.R., Morrison, M.L., Strickland, M.D., Thresher, R. (2007) Impacts of Wind Energy Facilities on Wildlife and Wildlife Habitat. Wildlife Society Technical Review 07-2, The Wildlife Society, Bethesda, MD, USA.

[10] Wilhelmsson, D.M. (2010) Greening Blue Energy: Identifying and Managing the Biodiversity Risks and Opportunities of Offshore Renewable Energy. IUCN.

[11] Wind Energy Siting Handbook, American Wind Energy Association. Available at: http://www.awea.org/Issues/Content.aspx?ItemNumber=5726 (last accessed 17 July 2014).

[12] Copes, R., Rideout, K. (2009) Wind Turbines and Health: A Review of Evidence. Ontario Agency for Health Protection and Promotion.

[13] Smedley, A.R., Webb, A.R., Wilkins, A.J. (2010) Potential of Wind Turbines to Elicit Seizures under Various Meteorological Conditions. *Epilepsia*, **51**(7), 1146–1151.

[14] Harding, G., Harding, P., Wilkins, A. (2008) Wind Turbines, Flicker, and Photosensitive Epilepsy: Characterizing the Flashing that May Precipitate Seizures and Optimizing Guidelines to Prevent Them. *Epilepsia*, **49**(6), 1095–1098.

[15] WindPRO 2.6 User Guide (2008) EMD International A/S.

[16] Gipe, P. (2002), *Wind Power in View: Energy Landscapes in a Crowded World*, Academic Press, San Diego, CA, USA.

[17] Seltenrich, N. (2014) Wind Turbines: A Different Breed of Noise? *Environmental Health Perspectives*, **122**, A20–A25; 10.1289/ehp.122-A20.

[18] IEC 61400-11 (2002) Ed.2 – Wind Turbine Generator System – Part 1: Acoustic Noise Measurement Techniques.

[19] American Wind Energy Association (AWEA) Standard (1989) Procedure for Measurement of Acoustic Emission from Wind Turbine Generator Systems, Tier I-2.1.

[20] International Energy Agency (IEA) (1994) Expert Group Study on Recommended Practices for Wind Turbine Testing and Evaluation: 4. Acoustic Noise Emission from Wind Turbines.

[21] Fégeant, O. (1999) On the Masking of Wind Turbine Noise by Ambient Noise. *Proceedings of the European Wind Energy Conference*, Nice, France, 1–5 March 1999.

[22] ISO 9613-2 (1996) Attenuation of Sound during Propagation Outdoors: Part 2. A General Method of Calculation.

[23] WindPRO 2.6 Manual (2008) EMD International A/S.

[24] Edenhofer, O., Pichs-Madruga, R., Sokona, Y., Seyboth, K., Matschoss, P., Kadner, S., Zwickel, T., Eickemeier, P., Hansen, G., Schlömer, S., von Stechow, C. (2011) *IPCC Special Report on Renewable Energy Sources and Climate Change Mitigation*. Prepared by Working Group III of the Intergovernmental Panel on Climate Change, Cambridge University Press, Cambridge, United Kingdom and New York, USA, 1075 pp.

[25] Zhou, L.M., Tian, Y.H., Roy, S.B., Thorncroft, C., Bosart, L.F., Hu, Y.L. (2012) Impacts of Wind Farms on Land Surface Temperature. *Nature Climate Change*, **2**, 539–543, doi:10.1038/nclimate1505.

[26] Wang, C., Prinn, R.G. (2010) Potential climatic impacts and reliability of very large-scale wind farms. *Atmospheric Chemistry and Physics*, **10**, 2053–2061.

[27] Wang, C., Prinn, R.G.. (2011) Potential Climatic Impacts and Reliability of Large-Scale Offshore Wind Farms. *Environmental Research Letters*, **6**, 025101 (6 pp.), doi:10.1088/1748-9326/6/2/025101.

[28] Emeis, S. (2013) *Wind Energy Meteorology*, Springer Verlag, Berlin and Heidelberg.

Appendix I

Frequently Used Equations

Name	Equation	Variables
Wind shear	$\alpha = \dfrac{\ln\left(V_2/V_1\right)}{\ln\left(Z_2/Z_1\right)}$	V_2 and V_1 are the mean wind speeds at heights of Z_2 and Z_1 respectively
Mean wind speed	$V_{ave} = \dfrac{1}{n}\displaystyle\sum_{i=1}^{n} V_i$ $V_{ave} = a\Gamma\left(1 + \dfrac{1}{k}\right)$ (Weibull) $V_{ave} = a\Gamma\left(\dfrac{3}{2}\right)$ (Rayleigh)	a is the Weibull scale parameter and k the slope parameter
Weibull cumulative distribution	$P(V_1 < V < V_2)$ $= \exp\left[-\left(\dfrac{V_1}{a}\right)^k\right] - \exp\left[-\left(\dfrac{V_2}{a}\right)^k\right]$	Same as above
50-year extreme wind speed	$x_{50} = \alpha y + \dfrac{\beta}{\alpha} + \alpha \ln 50$ $= x_1 + \alpha \ln 50 = x_T + \alpha \ln\left(\dfrac{50}{T}\right)$ $x_T = \alpha \cdot \max\{y_1, y_2 \cdots y_j \cdots y_n\} + \dfrac{\beta}{\alpha}$	α and β are the two parameters in the Gumbel distribution
Ideal gas law	$P = \rho R T$	ρ is air density, T is the temperature in K, $R = 287.05$ J/kg K,

Wind Resource Assessment and Micro-siting, Science and Engineering, First Edition. Matthew Huaiquan Zhang.
© 2015 John Wiley & Sons, Ltd. Published 2015 by John Wiley & Sons, Ltd.

Name	Equation	Variables
Air density	$\rho(z) = \dfrac{P_r}{RT} \exp\left(-\dfrac{g\,(z - z_r)}{RT}\right)$ or $$\rho(z) = \dfrac{352.9886}{T} \exp\left(-0.034163\dfrac{z}{T}\right)$$ where $T = T_r - 6.5(z - z_r)$	P_r and T_r are the air pressure and temperature at a reference level z_r, and z is the targeted height, T is the targeted temperature

Appendix II

IEC Classification of Wind Turbines

Wind turbine class		I	II	III	S
V_{ave} (m/s)		10	8.5	7.5	
V_{ref} (m/s)		50	42.5	37.5	
$V_{50,gust}$ (m/s)		70	59.5	52.5	User defined
I_{ref}	A	0.16			
	B	0.14			
	C	0.12			

In the table (IEC61400-1: 2005):

1. Rayleigh distribution is assumed, i.e. $k = 2$.
2. V_{ave} is the annual mean wind speed at hub height; V_{ref} is the 50-year extreme wind speed over 10 minutes; $V_{50,gust}$ is the 50-year extreme gust over 3 seconds; I_{ref} is the mean turbulence intensity at 15 m/s.
3. A, B and C are the categories of higher, medium and lower turbulence intensity characteristics respectively.

Wind Resource Assessment and Micro-siting, Science and Engineering, First Edition. Matthew Huaiquan Zhang.
© 2015 John Wiley & Sons, Ltd. Published 2015 by John Wiley & Sons, Ltd.

Appendix III

Climate Condition Survey for a Wind Farm

	Average climate parameters of wind turbines	Symbol	The representative met mast or the average conditions for all turbines	Unit
	Normal year			
	Measurement period	–	to	yymmdd
1	Measurement height	m.a.g.l.		m
2	Weibull scale parameter	A		m/s
3	Weibull shape parameter	k		–
4	Mean wind speed	v_{ave}		m/s
5	Mean ambient turbulence intensity at 15 m/s	I_{15}		%
6	Standard deviation of I_{15}	σ_{15}		%
7	Mean air density	ρ_{ave}		kg/m^3
8	Maximum air density	ρ_{max}		kg/m^3
9	Minimum air density	ρ_{min}		kg/m^3
10	Exponential wind shear	α		–
11	Maximum inflow angle		$\pm$	°
	50-year extreme wind conditions			
	Measurement period	–	to	yyyymm
12	Measurement height	m.a.g.l		m
13	10 min extreme mean wind speed	$V_{50,10min}$		m/s
14	3 s extreme gust wind speed	$V_{50,3s}$		m/s

Wind Resource Assessment and Micro-siting, Science and Engineering, First Edition. Matthew Huaiquan Zhang.
© 2015 John Wiley & Sons, Ltd. Published 2015 by John Wiley & Sons, Ltd.

	Average climate parameters of wind turbines	Symbol	The representative met mast or the average conditions for all turbines	Unit
	Environmental conditions			
	Measurement period	–	to	yyyymm
15	Mean annual ambient temperature	T_{50}		°C
16	1% percentile, minimum temperature	T_1		°C
17	99% percentile, maximum temperature	T_{99}		°C
18	Days of temperature > 40 °C (turbine specific)	–		Days/year
19	Days of temperature < −20 °C (turbine specific)	–		Days/year
20	Days of ice build-up	–		Days/year
21	Mean relative humidity	RH		%
22	Salt spray present	–	□ yes □ no	–
23	Lightening density	–		Strikes/km²/year

Edited from the climatic condition form of Vestas Wind System A/S.
Note: The air density and ambient temperature range should be estimated at the average wind turbine hub height.

III.1   Calculating the Ambient Temperature Range

Assuming a normal distribution individually for an average high temperature and an average low temperature, we have 1% percentile minimum temperature, T_1 and 1% percentile maximum temperature (i.e. 99% percentile low tempeture, T_{99}) as the following:

$$T_1 = T_L - 2.236\sigma_L$$
$$T_{99} = T_H + 2.236\sigma_H \qquad \text{(AIII.1)}$$

where T_H is the highest monthly average temperature, T_L is the lowest monthly average temperature, and σ_L and σ_H are the corresponding standard deviations.

Consequently, the 50-year extreme low and high temperatures are not considered given the 20-year lifespan of wind turbines. Using data from http://weatherbase.com/ as an example (see Figure AIII.1), we find that the highest monthly average temperature occurs in July, and $T_H = 25$ °C. The average high temperature in July is 31 °C, which gives us a standard deviation of 6 °C. Thus, we have the 1% percentile maximum ambient temperature:

$$T_{99} = 25 + 2.236 \times 6 = 39\ ^\circ\text{C}$$

TEMPERATURE

Average Temperature   Years on Record: 7

	ANNUAL	JAN	FEB	MAR	APR	MAY	JUN	JUL	AUG	SEP	OCT	NOV	DEC
C	9	−9	−5	2	10	18	23	25	23	18	10	---	−6

Average High Temperature   Years on Record: 7

	ANNUAL	JAN	FEB	MAR	APR	MAY	JUN	JUL	AUG	SEP	OCT	NOV	DEC
C	16	−2	2	9	18	26	29	31	29	25	17	7	---

Average Low Temperature   Years on Record: 7

	ANNUAL	JAN	FEB	MAR	APR	MAY	JUN	JUL	AUG	SEP	OCT	NOV	DEC
C	2	−16	−12	−5	3	11	17	20	18	11	3	−6	−13

Highest Recorded Temperature   Years on Record: 7

	ANNUAL	JAN	FEB	MAR	APR	MAY	JUN	JUL	AUG	SEP	OCT	NOV	DEC
C	40	10	18	23	32	37	40	37	37	33	30	22	12

Lowest Recorded Temperature   Years on Record: 7

	ANNUAL	JAN	FEB	MAR	APR	MAY	JUN	JUL	AUG	SEP	OCT	NOV	DEC
C	−25	−25	−25	---	−13	−1	7	12	11	−1	−10	−22	−25

Figure AIII.1   Example page from weatherbase.com

Similarly, we can achieve the 1% percentile minimum ambient temperature by applying the lowest monthly temperature of −9 °C (January) and the standard deviation of −7 °C, that is

$$T_1 = -9 - 2.326 \times 7 = -25.3 \,^\circ\text{C}$$

Therefore, the ambient temperature range for this particular site is from −25.3 to 39 °C.

Appendix IV

Useful Websites and Database

Knowledge

http://www.windpower.org/en/
Extensive knowledgebase about wind energy and technology

Weather Stations

http://www.weatherbase.com
Climate averages and forecasts for 41 997 cities worldwide

Magnetic Declination Calculator

http://www.ngdc.noaa.gov/geomag-web/#declination
Magnetic field calculators, estimating the value of local magnetic declination

Wind Resource and GIS Data

3Tier	http://www.vortexfdc.com/
Vortex	http://www.3tier.com/en/
WINDExchange	http://apps2.eere.energy.gov/wind/ windexchange/windmaps/
NREL	http://www.nrel.gov/gis/
SRTM Global Elevation Data	http://www2.jpl.nasa.gov/srtm/
WorldWind Global Elevation Data and Satellite Imagery	http://www.worldwindcentral.com/wiki/ Main_page

Wind Energy Associations

European Wind Energy Association (EWEA)	http://www.ewea.org/
American Wind Energy Association (AWEA)	http://www.awea.org/
World Wind Energy Association (WWEA)	http://www.wwindea.org/home/index.php
Global Wind Energy Council (GWEC)	http://www.gwec.net/

Wind Resource Assessment and Micro-siting, Science and Engineering, First Edition. Matthew Huaiquan Zhang.
© 2015 John Wiley & Sons, Ltd. Published 2015 by John Wiley & Sons, Ltd.

Index